Power Systems

Electrical power has been the technological foundation of industrial societies for many years. Although the systems designed to provide and apply electrical energy have reached a high degree of maturity, unforeseen problems are constantly encountered, necessitating the design of more efficient and reliable systems based on novel technologies. The book series Power Systems is aimed at providing detailed, accurate and sound technical information about these new developments in electrical power engineering. It includes topics on power generation, storage and transmission as well as electrical machines. The monographs and advanced textbooks in this series address researchers, lecturers, industrial engineers and senior students in electrical engineering.
** Power Systems is indexed in Scopus**

More information about this series at http://www.springer.com/series/4622

Morteza Zare Oskouei
Behnam Mohammadi-Ivatloo

Integration of Renewable Energy Sources Into the Power Grid Through PowerFactory

Morteza Zare Oskouei
Faculty of Electrical and Computer Engineering
University of Tabriz
Tabriz, Iran

Behnam Mohammadi-Ivatloo
Faculty of Electrical and Computer Engineering
University of Tabriz
Tabriz, Iran

Department of Energy Technology
Aalborg University
Aalborg, Denmark

ISSN 1612-1287 ISSN 1860-4676 (electronic)
Power Systems
ISBN 978-3-030-44378-8 ISBN 978-3-030-44376-4 (eBook)
https://doi.org/10.1007/978-3-030-44376-4

This Springer imprint is published by the registered company Springer Nature Switzerland AG
The registered company address is: Gewerbestrasse 11, 6330 Cham, Switzerland

Preface

Over the last two decades, the use of renewable energy sources has increased in power systems due to environmental concerns, nuclear power plant problems, and energy security risks. Renewable energies are sustainable energy resources and have environmental benefits. However, there are some substantial challenges when integrating them into existing power systems. This book addresses important requirements in the industrial and academic sectors to achieve a comprehensive resource for modeling, exploitation, control, as well as addressing the common challenges of renewable-integrated power grids. In addition, this work fully covers all technical issues and professional fields of power grid operation in the presence of renewable energy sources. DIgSILENT PowerFactory, as the most preferential tool in research and engineering problems, has been used in several studies and practical cases. In most countries, the results of technical analysis using this software is required to approve the installation of new renewable energy power plant. Thus, the availability of developed analysis and the proper descriptions in the form of the DIgSILENT PowerFactory shall remove barriers on extending studies and duplication of practical projects. Meanwhile, the application and study cases are selected with as many real implications.

This book is organized into seven chapters and presents a broad range of applied analysis in the renewable-integrated power systems studies as follows:

- Chapter 1 identifies the required actions for the techno-economic assessment of renewable energy sources. It also introduces the most common existed standards for the proper use of renewable energy sources.
- Chapter 2 presents the required tips for modeling key elements of the power grid in DIgSILENT PowerFactory.
- Chapter 3 gives the essential tips and instructions on how to model renewable energy sources within the framework of power systems by using DIgSILENT PowerFactory.
- Chapter 4 covers the power quality issues. In addition, the harmonic analysis based on renewable energy sources is explained in the framework of the DIgSILENT PowerFactory.

- Chapter 5 addresses the prevalent electrical challenges resulting from the misuse of renewable energy sources during the operation of power systems.
- Chapter 6 deals with the examination and calculation of reliability indices in the presence of renewable energy sources.
- Chapter 7 analyzes the technical parameters in the real grid-connected photovoltaic systems. In this chapter, the authors attempt to better illustrate the impact of renewable energy sources on power systems using a real distribution network.

Power engineers, educators, power system operators, researchers, and system designers can be among the specialized audience of this book. This book is intended for readers to familiarize themselves with the renewable-integrated power grids operation issues. In addition, the proposed framework of this book enables industry practitioners seeking to master the subject through self-education and training. It is also very useful for supporting teaching postgraduate courses in the field of operation of renewable sources.

Finally, this book is the result of the authors' research on the challenges of the renewable-integrated power grids. However, there is always space for improvement in all aspects. Hence, any comments and suggestions from the readers are welcome to improve the content presented in this book. Please share them with the authors via smart@tabrizu.ac.ir.

Tabriz, Iran — Morteza Zare Oskouei
Tabriz, Iran — Behnam Mohammadi-Ivatloo
December 2019

Acknowledgments

First and foremost, I would like to thank God. You have given me the power to believe in my passion and pursue my dreams. I could never have done this without the faith I have in you, the Almighty.

Most importantly, I would like to thank my parents and brother, Mojtaba, for the support and encouragement they provided me throughout this period is invaluable.

Morteza Zare Oskouei

Contents

Chapter 1
Introduction to Techno-Economic Assessment of Renewable Energy Sources

Over the past two decades, the International Energy Agency (IEA) has focused on encouraging various countries to invest in multi-source power generation systems, including renewable energy sources (RES) and associated power grids. Accordingly, it is imperative that all decision-makers and national experts master the technical and economic requirements regarding the integration of renewable sources into power systems. In general, the assessment and feasibility studies of renewable energy installations are carried out by different standards in four steps [1, 2]:

Step 1) Initial assessment of on-site renewable energy;
Step 2) Determination of most appropriate scale of renewable sources;
Step 3) Economic assessment of the level of risk, capital costs, payment and return on investment;
Step 4) Technical assessment of planning constraints that may affect the viability of the project.

All of the four critical planning steps mentioned above have a special role in developing RES worldwide. However, the goal of writing this book is to provide the key technical tools needed for the exploitation of RES (**Step 4**). By reading this book, the researchers and decision-makers will be able to assess the necessary technical requirements for the grid connection of the RES in the main power grids.

In this regard, DIgSILENT PowerFactory is widely used as one of the most powerful introduced software to evaluate the performance of power systems under various conditions, especially in the presence of RES. Hence, in the next chapters, all the required technical analysis for grid connection of RES sources are studied using DIgSILENT PowerFactory in full detail.

M. Zare Oskouei, B. Mohammadi-Ivatloo, *Integration of Renewable Energy Sources Into the Power Grid Through PowerFactory*, Power Systems,
https://doi.org/10.1007/978-3-030-44376-4_1

1.1 Renewable Energy Site Assessment

Identifying suitable locations as well as estimating the potential of renewable energy sources (RES) at each site is an essential pre-requisite for utilizing these sources, as well as increasing the capacity of clean and secure sources for electricity generation. All renewable sources depend on environmental factors and the availability of each of them varies in different regions of the globe. To better illustrate this fact, as an example, Figs. 1.1 and 1.2 show the variation of photovoltaic (PV) power and wind energy in different regions of Iran based on the amount of solar radiation and wind speed.

RES assessment techniques can be performed at various levels (e.g., strategic level and the more granular level) for many different types of users to determine the amount of energy harvested from each region at different times of the year. The required processes to assess renewable sources can be carried out in three distinct levels. These levels, along with the layers of each level, are shown in Fig. 1.3 [5]. The first level examines a wide range of RES, which can be useful in mapping broad sources availability. The second level assessment is conducted at the national level

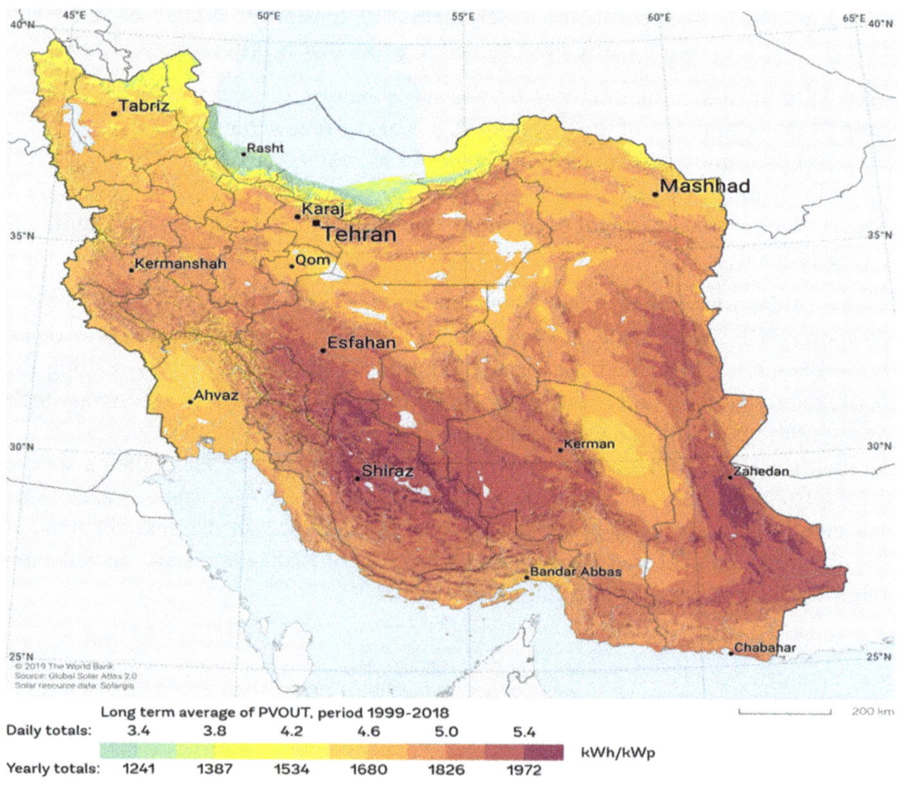

Fig. 1.1 Photovoltaic power potential in different regions of Iran [3]

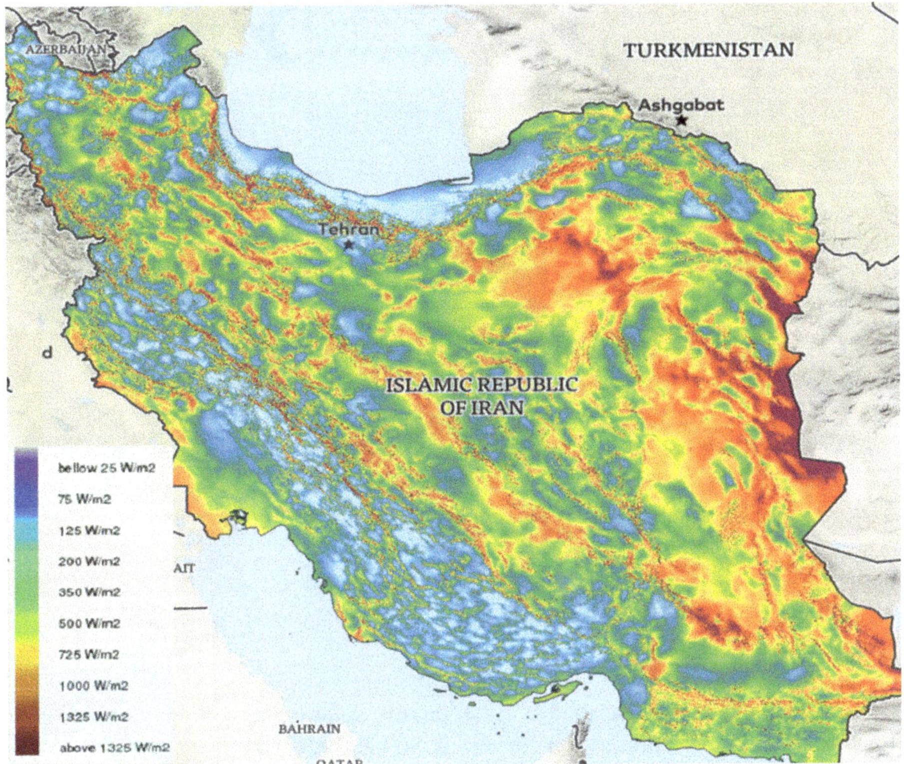

Fig. 1.2 Wind power potential at 50 m in different regions of Iran [4]

and provides more acceptable outputs than the first level by adding validation indices. For this reason, the outputs from the second level can be used for strategic planning and preliminary site identification purposes. Unlike the previous two levels, the third level assessments are based on environmental-technical-economic analysis. At this level, decision-makers examine the commercial viability and the profitability of each project at different sites.

Energy Sector Management Assistance Program (ESMAP) provides overall templates for assessing and mapping RES to support decision-makers and national experts [6]. The following is a summary of the presented templates for assessing and mapping solar, wind, and biomass sources.

1.1.1 Solar Energy

The solar system relying on various technologies captures the sun's energy and turns it into thermal or electrical energy. Solar energy is an important source of renewable energy and can be harnessed for a variety of activities. The technologies

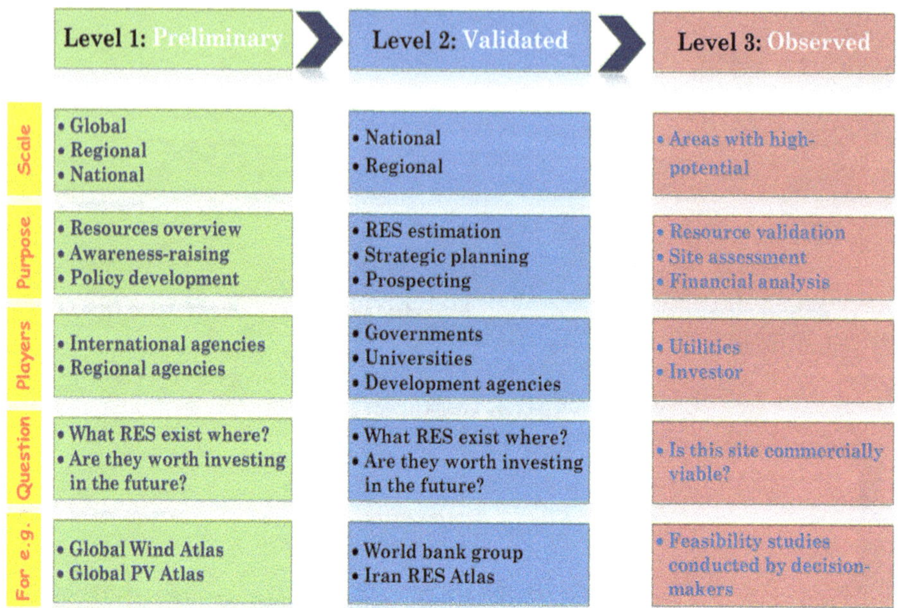

Fig. 1.3 Three levels for RES assessment

of the solar system are described as either passive solar or active solar, depending on how solar energy is captured and converted. So far, three ways have been introduced to use solar energy as the active mode: (1) PV systems, (2) concentrating solar power, and (3) solar heating and cooling [7, 8]. Solar resource assessment is strongly dependent on the accuracy of the collected data from the project site. The three main phases to assess solar resources are illustrated in Fig. 1.4 [6].

1.1.2 Wind Energy

Wind energy is a popular and sustainable alternative to fossil fuels. In today's society, the wind energy industry plays a special role in reducing environmental pollution, tackling climate changes, and increasing job opportunities [9]. Wind energy can have many uses, but nowadays, it is mostly used to generate electricity at different levels. The amount of energy that can be harvested from a wind turbine depends on the size of the turbine, the project site, and the length of the blades [10]. Due to the role of land cover, topography, and obstacles within any region on wind resource assessment, the mapping of this source is extremely complex [11]. However, the general template proposed by ESMAP for assessing wind resources is presented in Fig. 1.5 [6].

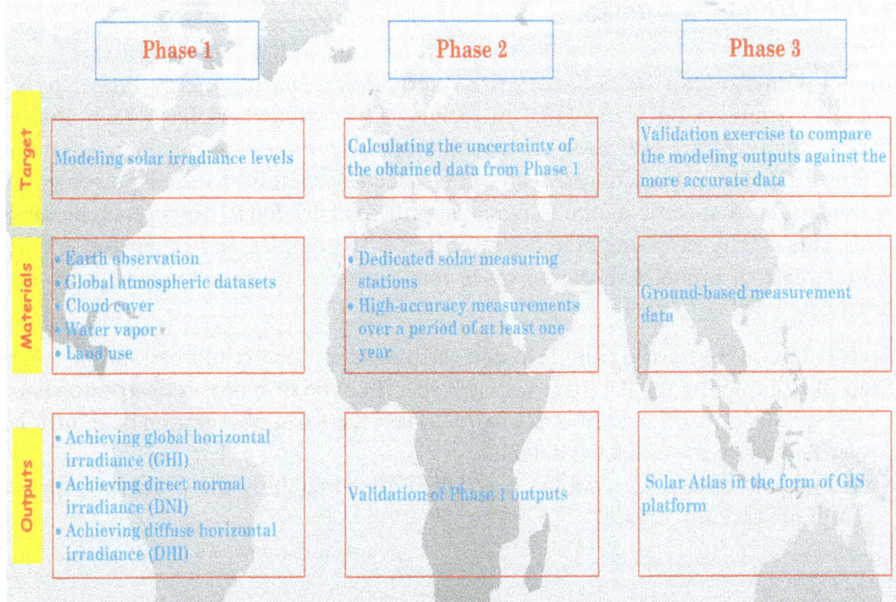

Fig. 1.4 The required main phases for the solar energy assessment

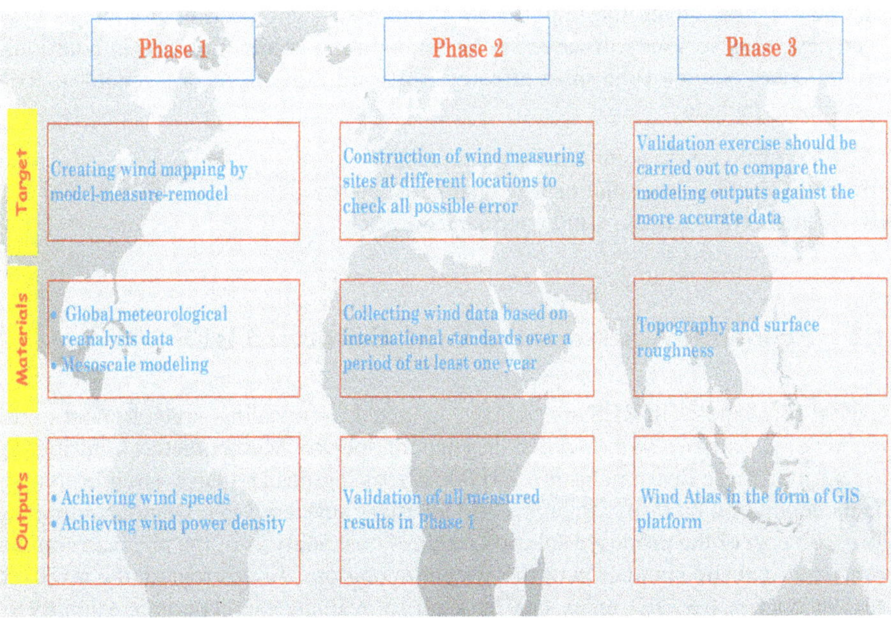

Fig. 1.5 The required phases for the wind energy assessment

1.1.3 Biomass Energy

Biomass energy can be collected from agricultural waste, energy crops, living organisms, and wood waste [12]. In recent years, researchers have been able to obtain various forms of hydrogen through biomass energy by using chemical interactions [13]. The extracted hydrogen can be used to generate electrical power and to fuel vehicles. A specific methodological approach is needed to assess each biomass type. This is why assessing biomass resources is so complex and expensive. A typical template of biomass resources assessment consists of three main steps as outlined below [6]:

Step 1) Use of the existing and foreseen earth plans to specify biomass-prone areas.
Step 2) Completing the database for each specific type of biomass using site visits, consultative events, and questionnaires. Then the earth observation data must be verified using the collected database.
Step 3) The biomass atlas should be created after matching the database with the earth observation data.

1.1.4 Commercialized Software Packages

As mentioned, RES assessment and mapping are very time-consuming and costly. In addition, decision-makers must take different actions in all fields to get highly accurate results. Given the importance of this issue, engineering companies have been designing various software to reduce computing volumes as well as additional capital costs. Some of the most effective software introduced in the field of RES assessment include:

I. **PVsyst**: to assess solar energy resources;
II. **PVSOL**: to assess solar energy resources;
III. **WindPRO**: to assess wind energy resources.

1.2 Economic Assessment of Grid-Connected RES

Capital investments in RES are tightly connected to financing strategies and to the market environments within which they would operate. Market factors contributing to return on these investments are variable across regions and have variable components depending on tariff structures and ancillary services [14]. Demonstrating the long-run cost of the proposed solution requires cost analysis of the physical components and realistic simulation of the system operations. In this regard, the levelized cost of energy (LCOE) index can be used to evaluate the economic viability of renewable sources projects. The LCOE is a valuable metric for determining the levelized cost of the energy generated by the different energy-producing technologies [15]. The main inputs for economic analysis of electrical energy-producing systems include investment cost, annual degradation rate, extended warranty, and annual operation and maintenance cost. The LCOE index could be declared mathematically by Eq. (1.1) [16].

$$\mathrm{LCOE} = \frac{\sum_{t=1}^{n}\left(I_t + \mathrm{OM}_t - \mathrm{PTC}_t - D_t + R_t\right)\frac{1}{\left(1+i\right)^t}}{\mathrm{IF}\sum_{t=0}^{n-1} P_t} \tag{1.1}$$

where:

t is the duration of the generation period (years);

i is the discount rate;

I_t is the investment made in year t ($);

OM_t is the operation and maintenance costs in year t ($);

PTC_t is the production tax credit ($);

D_t is the depreciation credit ($);

R_t is the land rents ($);

P_t is the electrical generation capacity in year t (kWh);

IF is the intermittence factor. Intermittency here means the fact that RES only produce electricity when nature cooperates. For example, PV systems only operate when the sun shines.

Typically, to calculate the LCOE index, the lifetime of the energy-producing technology is assumed to be 20 years. Given the importance of the financial issues in each project, the general concept of the LCOE index is shown in Fig. 1.6.

1.2.1 Feed-in Tariff Policy

The high costs of research and development (R&D) in the green energies sector make them economically unjustifiable. In recent years, leading countries in the field of green energy have used various schemes to accelerate the development of renewable resources. The Feed-in-Tariff (FiT) policy is one of the most prominent schemes in the RES development process [17]. In the form of FiT schemes, decision-makers enter into long-term contracts with renewable energy producers to overcome investors' financial worries. The first template to implement the FiT scheme was presented after the US president signed the National Energy Act in 1978 [18]. The three main advantages for renewable energy investors in these schemes are guaranteed power network access, long-term power purchase agreements, and cost-based purchase prices [19]. In general, these schemes are implemented in market-dependent and market-independent FiT policies [20]. Most countries focus on market-dependent FiT policies. Market-dependent FiT schemes require that RES investors

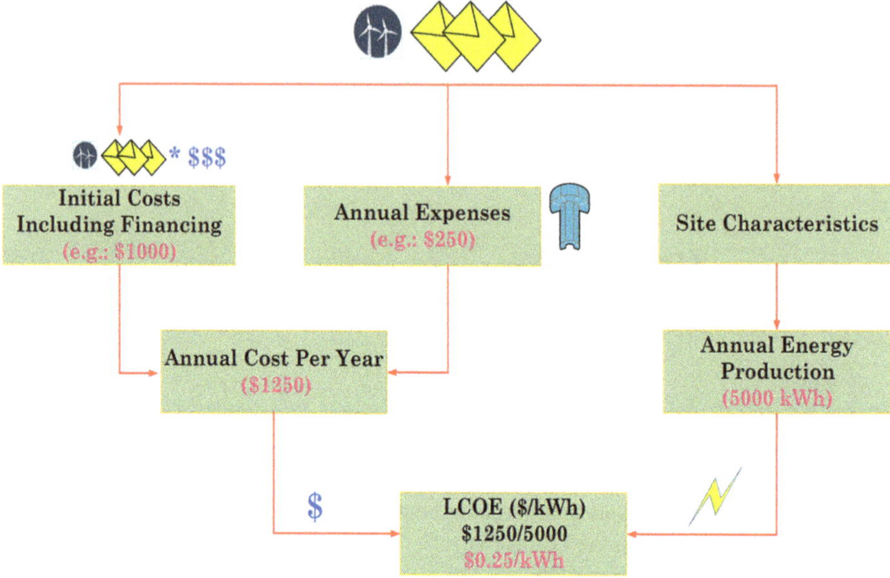

Fig. 1.6 Conceptual architecture of the LCOE index

provide their electricity to the electricity market, effectively competing with other investors to meet market demand, and then they receive a premium above the spot market price for the electricity sold.

The FiT regulations for different types of RES in various countries are summarized in Tables 1.1, 1.2, 1.3, 1.4, 1.5, and 1.6. The exchange rates are provided at the end of tables as bold items in 2019. According to these tables, most governments provide special financial incentives for PV systems. For this reason, PV sources have expanded dramatically throughout the world in recent years.

1.3 Technical Requirements for Connecting RES to Power Grids

This section reviews the technical requirements for connecting renewable resources to power grids according to the IEEE 1547 standard and the common guidelines in different countries as well as provides recommendations for further improvement. The main targets of the power system operators and planners are directly related to the time index [27]. For example, power quality issues, as well as voltage and frequency stability, are targets of concern on short time scales (from milliseconds to minutes). Supply and demand balance are other targets for the power system operators that must be satisfied on medium time scales (from minutes to hours). The generation and transmission capacity as the principle indices on longer time scales (from weeks to seasons), must be able to meet electrical demand in all parts of the power grid over the whole year. On the other hand, the investigation of these indices

Table 1.1 Feed-in tariffs in Malaysia [21]

Renewable energy	FiT rate Effective period (year)	FiT rate $/kWh[a]	Annual digression (%)
Biomass			
<10 MW	16	0.075	0.5
10–20 MW	16	0.069	0.5
20–30 MW	16	0.064	0.5
Small hydro			
<10 MW	21	0.057	0
10–30 MW	21	0.055	0
Photovoltaic			
<4 kWp (kilowatt peak)	21	0.29	8
4–24 kWp	21	0.288	8
24–72 kWp	21	0.283	20
72 kWp–1 MWp	21	0.273	20
1–10 MWp	21	0.228	20
10–30 MWp	21	0.204	20

[a]The rate of exchange for FiT index is **1 RM** (Malaysian Currency) = **0.24 $**

Table 1.2 Feed-in tariffs in Iran [22]

Renewable energy	FiT rate Effective period (year)	FiT rate $/kWh[a]	Description
Biomass			The tariffs mentioned for all power plants are multiplied by **0.7** after the first ten years of the contract.
Landfill	20	0.063	
The anaerobic digestion of manure, sewage, and agriculture	20	0.081	
Incineration and waste gas storage	20	0.087	
Wind			
<1 MW	20	0.133	
1–50 MW	20	0.098	
>50 MW	20	0.078	
Photovoltaic			
<20 kW	20	0.188	
20–100 kW	20	0.165	
100 kW–10 MW	20	0.116	
10–30 MW	20	0.092	
>30 MW	20	0.076	

[a]The rate of exchange for FiT index is **1 IRRs** (Iranian Currency) = **2.38×10^{-5} $**

in the presence of active components such as RES is crucial to maintain the stability of the power grid [28]. High penetration of various RES still leads to many technical challenges on power grids because parts of the power grid change from consumption mode to supply mode. The overall behavior of the power grids with variable

Table 1.3 Feed-in tariffs in Japan [23]

Renewable energy	FiT rate	
	Effective period (year)	$/kWh[a]
Biomass		
100 kW–5 MW	16	0.193
5–20 MW	16	0.174
>20 MW	16	0.156
Wind		
<1 MW	20	0.22
1–10 MW	20	0.165
>10 MW	20	0.11
Photovoltaic		
Residential installations	20	0.239
Nonresidential installation	20	0.211

[a]The rate of exchange for FiT index is **1 Yen** (Japanese Currency) = **0.0092 $**

Table 1.4 Feed-in tariffs in Michigan in the United States [24]

Renewable energy	FiT rate	
	Effective period (year)	$/kWh
Biomass		
<150 kW	20	0.145
150–500 kW	20	0.125
500 kW–5 MW	20	0.115
5–20 MW	20	0.105
Wind		
<700 kWh/m²/year	20	0.105
700–1100 kWh/m²/year	20	0.095
>1100 kWh/m²/year	20	0.08
1000 sq. ft. swept area	20	0.25
Photovoltaic		
<30 kW	20	0.71
30–100 kW	20	0.68
>100 kW	20	0.67
Rooftop < 30 kW	20	0.65
Rooftop 30–100 kW	20	0.62
Rooftop > 100 kW	20	0.61
Ground mounted	20	0.50
Hydro-power		
<500 kW	20	0.1
500 kW–10 MW	20	0.085
10–20 MW	20	0.065
Geothermal		
<5 MW	20	0.19
5–10 MW	20	0.18
10–20 MW	20	0.115
>20 MW	20	0.09

renewable generation differs from the dispatchable resources. Hence, power grid operators have been forced to instantly react to the locally growing RES penetration rates with modified grid planning and operation processes [29].

To tackle these challenges, some countries have provided a strict regulatory framework that considered the technical tools for power system operators to cope with the rising installation rates in a technically effective manner. In this regard, many attractive research fields were conducted on regional and national levels that aimed to examine grid operation and grid planning processes to improve the technical integration of RES in power grids. In reality, the power system planners need to examine all aspects of variable renewable sources integration, as their actual execution will in large part depend on the energy markets and the projected penetration level of RES. The following are the main technical requirements for integrating RES into modern power grids [30–32]:

I. **Voltage and reactive power control:**

This issue is closely related to the extent of the influence of renewable sources at the point of common coupling (PCC). According to various standards, the allowable deviation from the nominal voltage should be within the range of ±5 to ±10% after utilizing renewable energy sources.

II. **Frequency and active power control:**

The use of intermittent RES in coordination with power grids leads to increase or decrease of the generated active power and will ultimately affect the frequency of the power system. According to the existing standards, the allowable deviation from the nominal frequency should be within the range of −5 to +3% after utilizing RES.

III. **Short-circuit power level:**

From the viewpoint of power grid stability, network planners should evaluate the role of RES when various faults occurring in different zones of the system. Improper use of RES as inverter-based resources can have detrimental effects on some electric systems. Hence, various indices, especially short-circuit power level must be calculated with and without renewable sources and required analyzes must be conducted to prevent critical conditions in the power system.

IV. **Power quality issues:**

The main issue of power quality is the analysis of various harmonic orders that distort the waveform and short-term fluctuations. Common indicators of the power quality issue in the presence of RES need to be evaluated and respected. The following international standards can be used to power quality analysis in the context of RES interconnection as they describe most of the technical requirements and limitations:

1. **IEEE P1547:** Distributed resources and electric power systems interconnection;
2. **IEEE 1433:** Power quality definitions;
3. **IEEE 519:** Harmonic control in electrical power systems;

Table 1.5 Feed-in tariffs in Germany [25]

Renewable energy	FiT rate ($/kWh[a])	Description
Biomass	0.063–0.147 (according to plant size)	The tariff level will be reduced to the actual market value, if: **(1)** technical requirements are not met; **(2)** the plant operator has switched between direct marketing of the electricity and the feed-in tariff without informing the grid operator; **(3)** a plant operator breaches the prohibition of multiple sale electricity from renewable energy sources and mine gas.
Wind		
Onshore	0.051–0.093 (according to duration of payment)	
Offshore until 2020	0.043–0.015 (according to duration of payment and scheme chosen by plant operator)	
Photovoltaic		
Specific building-mounted systems (e.g. roofs, facades, noise barriers)	0.098-0.139	
Hydro-power	0.038–0.136 (depending on plant size and date of commissioning)	
Geothermal energy	0.277	

[a]The rate of exchange for FiT index is **1 € = 1.11 $**

Table 1.6 Feed-in tariffs in Turkey [26]

Renewable energy	FiT rate ($/kWh)	Local-content bonus ($/kWh)	Description
Biomass	0.133	0.004–0.018	(1) In general, the maximum installed capacity of a solar system should be less than 50 MW; (2) the feed-in tariff is limited to 10 years. The bonus tariff for local-content support is limited to the first 5 years of operation; (3) the license of production for RES is granted for 30 years in the form of FiT schemes.
Wind			
Onshore and offshore	0.073	0.006–0.037	
Photovoltaic			
PV system	0.133	0.006–0.067	
Concentrating solar power (CSP)	0.133	0.006–0.092	
Hydro-power	0.073	0.009–0.02	
Geothermal energy	0.105	0.007–0.027	

4. **IEEE 1159:** Monitoring electric power quality;
5. **IEEE SCC-22:** Power quality standards coordinating committee;
6. **IEEE P519A:** Guide for applying harmonic limits on power systems;
7. **IEEE P1564:** Voltage sag indices;
8. **IEEE 1346:** Power system compatibility with process equipment.

V. **Congestion management:**

The power grid may not be able to accommodate new RES capacity at the PCC and may constrain the path of power delivery from the point of RES connection to subscriber sectors. Power grid congestion may occur due to the deployment of new renewable generation capacity in various zones of the grid. The power system planners should be conducted a comprehensive analysis to identify weak components of the infrastructure that may cause congestion.

1.3.1 Overview of the IEEE 1547 Standard

The IEEE standard 1547 is an essential document to guide decision-makers to properly communicate between distributed energy resources and the power grid. This standard has helped to modernize the power grids infrastructure by offering a strong foundation for integrating RES technologies as well as energy storage systems. The IEEE 1547 standard presents mandatory technical requirements relevant to the interoperability performance, maintenance and security considerations, as well as essential equipment and details of their operation. The U.S. Department of Energy (DOE) and National Renewable Energy Laboratory (NREL) have played a special role in developing the IEEE 1547 standard, and nowadays it is used as the only American National Standard in the distribution sector [33]. For the interconnection of the RES sources, the following are some of the most important standards in the form of the IEEE 1547 standard:

1. **IEEE 1547.1**: This part of the IEEE 1547 standard is applied to test the conformance of equipment used in distributed energy resources when connecting to power grids.
2. **IEEE 1547.2**: This standard is used as an application guide for the IEEE 1547 standard.
3. **IEEE 1547.3**: This part of the IEEE 1547 standard is used as a special guide for monitoring, control, and information exchange of distributed energy resources interconnected with power grids.
4. **IEEE 1547.4**: This standard is used as a unique guide for operation, design, and integration of distributed energy resources in islanding mode with power grids.
5. **IEEE 1547.6**: This part of the IEEE 1547 standard is used as the recommended practice for interconnecting distributed energy resources with distribution secondary networks.

6. **IEEE P1547.7**: This part of the IEEE 1547 standard is used as a practical guide for evaluating the impact of distributed energy resources on power grids.
7. **IEEE P1547.8**: This standard is applied to establish supplemental procedures and novel methods for extending the IEEE 1547 standard.

Overall, the developed version of the IEEE 1547 standard is addressing distribution-level connected distributed energy resources, which includes:

- Renewable generation units and storage technologies
- High penetration of distributed generation
- Uncertainty and intermittency of RES
- Advanced characteristics of both RES and modern grid components
- Requirements for considering evaluations of resiliency and reliability of RES-grid interconnections
- Incorporation of advanced testing and assessment methods such as enhanced modeling and simulation requirements
- Essential requirements to create mutual coordination between transmission and distribution networks
- Interoperability and smart devices integration to enhance network control level
- Demand response programs and dispatchable load effects

1.3.2 Overview of the Common Guidelines

In this part, we intend to review some guidelines for RES connected to the power grids in Malaysia, Indonesia, and Australia. The reports presented in this part summarize the issues of the technical and regulatory framework for connecting RES into the power grids in each country.

1. **First Case Study: Malaysia**

The following information is issued by Tenaga Nasional Berhad (TNB) company [34]. The provided guidelines to determine the operating voltage range of RES and distributed generation (DG) in normal and critical conditions are in accordance with Table 1.7. The overall guidelines to tackle the total harmonic distortion (THD) issue at the PCC are in accordance with Table 1.8.

The general guidelines to determine the maximum permissible installed capacity of the PV systems at various points are as follows:

- **For a single-phase low voltage (LV) feeder**: maximum permitted PV capacity is 54 kW to prevent voltage limit violation.
- **For a 415 V pillar feeder**: (1) 90% of the transformer capacity, (2) subject to each individual connection limited to 180 kW/250 A.

2. **Second case study: Indonesia**

The following information is collected according to the document guideline for connecting renewable energy generation plants (REGP) to PLN's distribution systems [35]. According to this guideline:

Table 1.7 Operating voltage range in normal and critical conditions

Nominal voltage	Guideline for DG		Guideline for RES	
	Normal mode	Critical mode	Normal mode	Critical mode
33 kV	±5%	±10%	±5%	Not specified
22 kV	±5%	±10%	±5%	
11 kV	±5%	±10%	±5%	
6.6 kV	Not specified	Not specified	±5%	
415 V	+5% and – 10%	±10%	Not specified	
400 V and 230 V	Not specified	Not specified	+10% and – 6%	

Table 1.8 Harmonic current distortion limits

Harmonic order	Current distortion limit (%)	Description
3–9	<4	Even harmonic shall be less than 25% of the lower odd harmonic limits listed.
11–15	<2	
17–21	<1.5	
23–33	<0.6	
>33	<0.5	

Table 1.9 Response to abnormal voltages

Voltage (V)	Maximum clearing time
$V < 50\%$	100 ms
$50\% \leq V < 85\%$	2 s
$85\% \leq V < 110\%$	Normal operation
$110\% \leq V \leq 135\%$	2 s
$V > 135\%$	100 ms

1. The renewable energy generation plants must disconnect from the power grid under abnormal voltage conditions. These conditions are expressed in Table 1.9.
2. The renewable energy generation plants must operate at their full output to maintain system frequency in the range 47.5–51 Hz. If the system frequency exceeds 50.5 Hz, an offline renewable energy generation plants should not be connected to the power grid.
3. The renewable energy generation plants must disconnect from the power grid under abnormal frequency conditions. These conditions are expressed in Table 1.10.
4. Common rules for dealing with the total harmonic distortion (THD) at the PCC are in accordance with Table 1.11. Also, even harmonic must be less than 25% of the lower odd harmonic limits listed in this table.

3. **Third Case Study: Australia**

The technical framework for RES interconnection in Australia is depicted in Fig. 1.7 [36–38]. Common standards in Australia for connecting RES to the distribution network are as follows:

Table 1.11 Harmonic current distortion limits

Harmonic order (H)	Maximum harmonic current distortion, % of I_n
$H < 11$	4
$11 \leq H < 17$	2
$17 \leq H < 23$	1.5
$23 \leq H < 35$	0.6
$H > 35$	0.3

Table 1.10 Response to abnormal frequency

Frequency (f)	Maximum clearing time
$f > 51$ Hz	10 s
$f < 47.5$ Hz	10 s

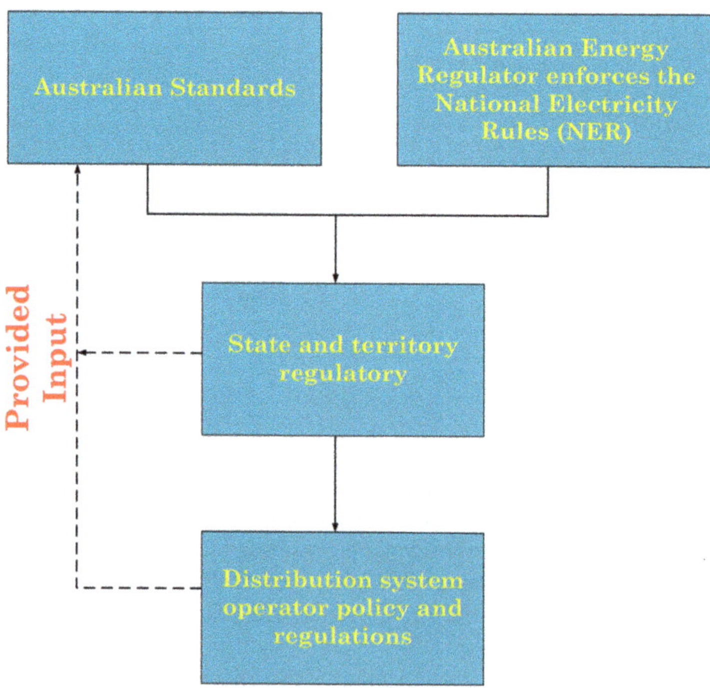

Fig. 1.7 Technical framework for RES interconnection in Australia

- **AS4777:** Grid connection of energy systems through the power electronic converters.
- **AS61000.3.100:** Steady-state voltage limits in the power grids.
- **AS3000:** The wiring guidelines.

Table 1.12 System size limits and voltage and frequency inverter set points

State	DNSP	PV system size limit	Voltage and frequency set points			
			f_{min} (Hz)	f_{max} (Hz)	V_{min} (V)	V_{max} (V)
NSW	AusGrid	10 kW per phase	48–50	50–52	200	260
NSW	Essential energy	10 kW per phase	48–50	50–52	200	260
NSW	Endeavour	30 kW per phase	48	52	190	260
NT	Power and water corporation	4.5 kW residential, 30 kVA for three-phase commercial	46	54	210	270
QLD	Energex	5 kW	48	52	–	255
SA	SA power networks	10 kW, 5 kW for SWER network	48	52	–	257
VIC	CitiPower	10 kW per phase, unless rural location	48.5	51.5	195	265
WA	Horizon power	10 kW per phase	46.5	53	190	265

Generally, the distribution network service providers (DNSPs) consider various regulations based on the requirements of the Australian standards and the state regulator. Most of these regulations provide the technical guidelines and the connection methods with which a PV system must comply. DNSPs in various states provide general guidelines for household subscribers to determine the PV system size limit as well as the voltage and frequency set points of the inverter. The system size limits and the set point values in the different states are summarized in Table 1.12.

References

1. Bianchini A, Gambuti M, Pellegrini M et al (2016) Performance analysis and economic assessment of different photovoltaic technologies based on experimental measurements. Renew Energy 85:1–11
2. Sgobba A, Meskell C (2019) On-site renewable electricity production and self consumption for manufacturing industry in Ireland: sensitivity to techno-economic conditions. J Clean Prod 207:894–907
3. Global solar atlas, [Online]. Available at: https://globalwindatlas.info/en/area/Islamic%20Republic%20of%20Iran?print=true/
4. Global wind atlas, [Online]. Available at: https://globalwindatlas.info/en/area/Islamic%20Republic%20of%20Iran?print=true/
5. Knight O (2016) Assessing and mapping renewable energy resources. Energy Sector Management Assistance Program (ESMAP) knowledge series
6. Energy Sector Management Assistance Program (ESMAP), [Online]. Available at: https://www.esmap.org/
7. Sampaio PGV, González MOA (2017) Photovoltaic solar energy: conceptual framework. Renew Sust Energ Rev 74:590–601
8. Kannan N, Vakeesan D (2016) Solar energy for future world: a review. Renew Sust Energ Rev 62:1092–1105

9. Kumar Y, Ringenberg J, Depuru SS et al (2016) Wind energy: trends and enabling technologies. Renew Sust Energ Rev 53:209–224
10. Alonzo B, Ringkjob HK, Jourdier B et al (2017) Modelling the variability of the wind energy resource on monthly and seasonal timescales. Renew Energy 113:1434–1446
11. Siyal SH, Mörtberg U, Mentis D et al (2015) Wind energy assessment considering geographic and environmental restrictions in Sweden: a GIS-based approach. Energy 83:447–461
12. Toklu E (2017) Biomass energy potential and utilization in Turkey. Renew Energy 107:235–244
13. Aslan A (2016) The causal relationship between biomass energy use and economic growth in the United States. Renew Sust Energ Rev 57:362–366
14. Lai CS, McCulloch MD (2017) Levelized cost of electricity for solar photovoltaic and electrical energy storage. Appl Energy 190:191–203
15. Nissen U, Harfst N (2019) Shortcomings of the traditional "levelized cost of energy" [LCOE] for the determination of grid parity. Energy 171:1009–1016
16. Bruck M, Sandborn P, Goudarzi N (2018) A Levelized cost of energy (LCOE) model for wind farms that include power purchase agreements (PPAs). Renew Energy 122:131–139
17. Pyrgou A, Kylili A, Fokaides PA (2016) The future of the feed-in tariff (FiT) scheme in Europe: the case of photovoltaics. Energy Policy 95:94–102
18. Public utility regulatory policies Act, United States (1978)
19. Sun P, Nie P (2015) A comparative study of feed-in tariff and renewable portfolio standard policy in renewable energy industry. Renew Energy 74:255–262
20. Couture T, Gagnon Y (2010) An analysis of feed-in tariff remuneration models: implications for renewable energy investment. Energy Policy 38:955–965
21. Wong SL, Ngadi N, Abdullah TAT et al (2015) Recent advances of feed-in tariff in Malaysia. Renew Sust Energ Rev 41:42–52
22. Renewable Energy and Energy Efficiency Organization (SATBA), [Online]. Available at: http://www.satba.gov.ir/en/guaranteed-Guaranteed-Renewable-Electricity-Purchase-Tariffs/
23. Ayoub N, Yuji N (2012) Governmental intervention approaches to promote renewable energies -special emphasis on Japanese feed-in tariff. Energy Policy 43:191–201
24. Rickerson W, Bennhold F, Bradbury J (2008) Feed-in tariffs and renewable energy in the USA – a policy update
25. Legal sources on renewable energy, [Online]. Available at: http://www.res-legal.eu/search-by-country/germany/
26. Legal sources on renewable energy, [Online]. Available at: http://www.res-legal.eu/search-by-country/turkey/
27. Behboodi S, Chassin DP, Djilali N et al (2017) Interconnection-wide hour-ahead scheduling in the presence of intermittent renewables and demand response: a surplus maximizing approach. Appl Energy 189:336–351
28. Liang X (2016) Emerging power quality challenges due to integration of renewable energy sources. IEEE Trans Ind Appl 53(2):855–866
29. Simoglou CK, Bakirtzis EA, Biskas PN et al (2016) Optimal operation of insular electricity grids under high RES penetration. Renew Energy 86:1308–1316
30. Kondziella H, Bruckner T (2016) Flexibility requirements of renewable energy based electricity systems – a review of research results and methodologies. Renew Sust Energ Rev 53:10–22
31. Weitemeyer S, Kleinhans D, Vogt T et al (2015) Integration of renewable energy sources in future power systems: the role of storage. Renew Energy 75:14–20
32. Anees AS (2012) Grid integration of renewable energy sources: Challenges, issues and possible solutions. Paper presented at the 5th India international conference on power electronics (IICPE), Delhi
33. Basso T, Chakraborty S, Hoke A et al (2015) IEEE 1547 Standards advancing grid modernization. Paper presented at the 42nd photovoltaic specialist conference (PVSC), New Orleans
34. Tenaga Nasional Berhad (TNB), [Online]. Available at: https://www.tnb.com.my/

35. PT. PLN (Persero); Guidelines for connecting renewable energy generation plants (REGP) to PLN's distribution system; July 2014
36. Australian standard (2007) AS/NZS 3000:2007 electrical installations
37. AEMC, Tasmanian frequency operating standard review, Sydney, 2008
38. AEMC reliability panel, application of frequency operating standards during periods of supply scarcity, Sydney, 2009

Chapter 2
Introducing Basic Tools in DIgSILENT PoweFactory

DIgSILENT PowerFactory is one of the most powerful software to technical analysis of the power grids and also determine the optimal operation of each element in the electrical network. This chapter presents the required tips for modeling key elements of the power grid in PowerFactory. It can be used to simulate and analyze different studies in the power grid.

2.1 Prevalent PowerFactory Capabilities for Power System Analysis

DIgSILENT PowerFactory is a computation software to assess electrical power transmission and distribution systems in order to achieve the main goals of operation and planning optimization of these systems. Hence, PowerFactory software plays an important role in conducting technical research by operators of the power system. Some of the analyses performed to evaluate the power system are as follows:

- Assessment of the power system's basic characteristics [1–4]: Evaluation of the power system under normal and abnormal conditions to maintain key parameters within the permissible range is one of the main tasks of the system's operators. PowerFactory provides users with all the required tools for analyzing power system in different modes. The most commonly used computing functions in this software include *Load Flow Analysis*, *Short-Circuit Analysis*, *Power Quality Analysis*, *Modal Analysis*, and *Reliability Analysis*.
- Application of probabilistic load flow calculation [5, 6]: This approach tries to handle the power system's uncertainties, including load increment and generation scheduling by considering Monte Carlo simulation and point estimate methods. By using this application, user can promote capabilities of the deterministic power flow.

M. Zare Oskouei, B. Mohammadi-Ivatloo, *Integration of Renewable Energy Sources Into the Power Grid Through PowerFactory*, Power Systems,
https://doi.org/10.1007/978-3-030-44376-4_2

- Modeling of transmission and distribution systems [7, 8]: In this study, the decision maker is trying to make optimal investments in new lines, feeders, and substations to cover the technical constraints.
- Active power control of wind farms [9]: The large-capacity wind farms bring various challenges to the power system operation. One of the major problems is related to wind power spillage due to the power grid constraints. Hence, decision makers are always trying to find the most suitable options for exploiting wind farms. This problem can be solved via real measurements of wind speeds and power generated by wind farms as well as utilizing the special tools of PowerFactory.
- Phasor measurement unit (PMU) allocation [10–13]: The challenges associated with the power system observability have been evaluated by relying on the optimal location of PMU.
- Assessment of transient stability [14–16]: Maintaining the stability of the power system is critical after sudden changes in load or unforeseen faults. Transient stability analysis examines how various methodologies are developed to reduce voltage and frequency fluctuations after different events. Electromagnetic transient (EMT) tool can be utilized to study the stability of the system in real-time.

2.2 Modeling Important Elements

Each power network consists of three main elements, power plant, transmission line, and electrical load. The following sub-sections present how to use these elements in the DIgSILENT PowerFactory software.

2.2.1 Power Plant Modeling

Power plant units or generating station is an industrial facility for the generation of electric power. In PowerFactory software, power plant units are modeled by synchronous generators. The required settings for proper operation of synchronous generator units are presented in detail in [17]. Considering the topics that will be discussed in the following chapters, the following sub-section describes how to utilize power plant units in the form of virtual power plant (VPP).

2.2.1.1 Virtual Power Plant

The VPP is a considerable solution for reliable supply of electricity in the power grid. The VPP integrates some decentralized power generation units, energy storage systems, and flexible loads via a standard intelligent control system. The sum of the generated power by the VPP should be equal to the concluded contract between

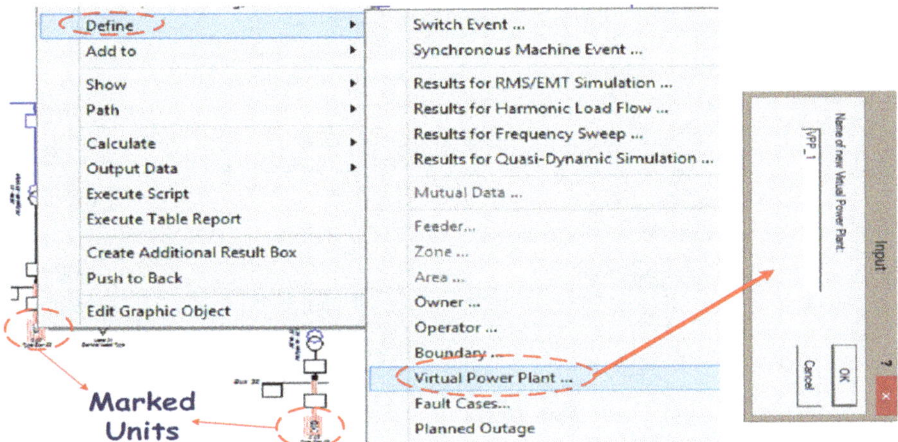

Fig. 2.1 Process of creating VPP

VPP owner and energy market operator. To define the VPP in PowerFactory, users must follow the steps below:

- *Hold down **Ctrl** and select the desired generators to create a VPP system*;
- *Right-click on the selected generators then choose "Virtual Power Plant" from "Define" option*;
- *Select the desired name for the VPP*;
- *In the pop-up window, "Active Power Setpoint" must be specified. This value must be provided by all units involved in the virtual system.*

These steps are shown in Fig. 2.1. The amount of imported power into the power network (active power setpoint) by any of the units involved in the virtual system, can be determined in two ways: (1) according to merit order, or (2) DIgSILENT programming language (DPL) command. According to Fig. 2.2, to use the merit order method, a specific amount must be assigned to each unit. In this method, to the extent that the assigned value to a unit is smaller, the amount of the injected active power to the power system by that unit will be greater. Some power plant units may be in must run state to maintain power grid stability, so they will not follow the rules of merit order. In addition, other generators can be added to the created virtual system. For this purpose, right-click on desired unit and use "*Add to*" option to add the desired unit to the virtual system.

2.2.2 *Line Modeling*

In the power system, power lines are utilized to transfer power at different voltage levels. These lines are known as the main element to communicate between the generation sector and distribution sector under various technologies. The design and

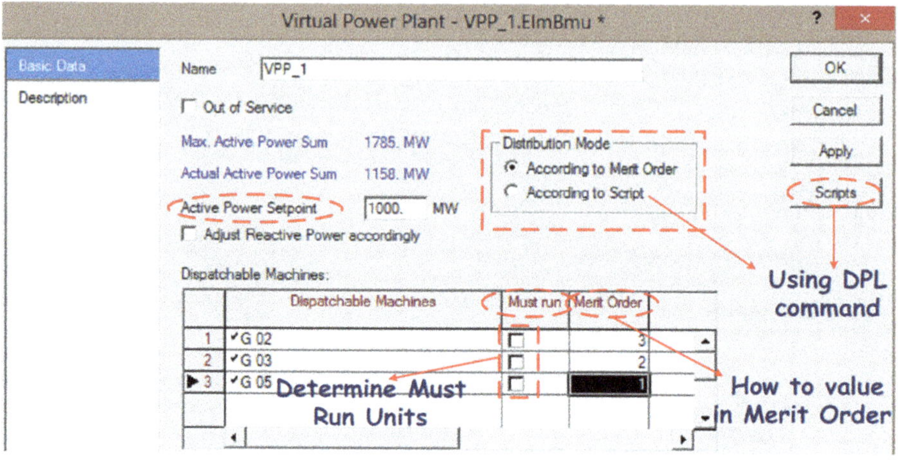

Fig. 2.2 Required settings to launch VPP systems

Table 2.1 Existing structures for line modeling

System	Phase technology	Structure	Type
AC	1 PH/neutral	Line	TypLne
		Cable	TypCabsys
	2 PH/neutral	Line	TypLne
		Cable	TypCabsys
	3 PH/neutral	Line	TypLne
		Cable	TypCabsys
		Tower	TypTow/TypGeo
DC	Unipolar	Line	TypLne

deployment of power lines in different parts of the power system depend on various factors such as the weather conditions, protection systems, and so on.

PowerFactory provides users with various capabilities for modeling power lines. All possible options for modeling power lines in the transmission and distribution sectors are presented in Table 2.1. The required details to model existing structures are presented below in separate sub-sections.

2.2.2.1 Single-Line Mode

This type is the most common mode for modeling power lines at different voltage ranges for both AC and DC systems. This model can be used in both overhead line and cable (under-ground) modes with different structural features to perform different simulations. All the properties and required parameters for modeling power lines in single-line mode are shown in Fig. 2.3. Aluminum, Copper, and Aldrey can be used as the conductor of the lines.

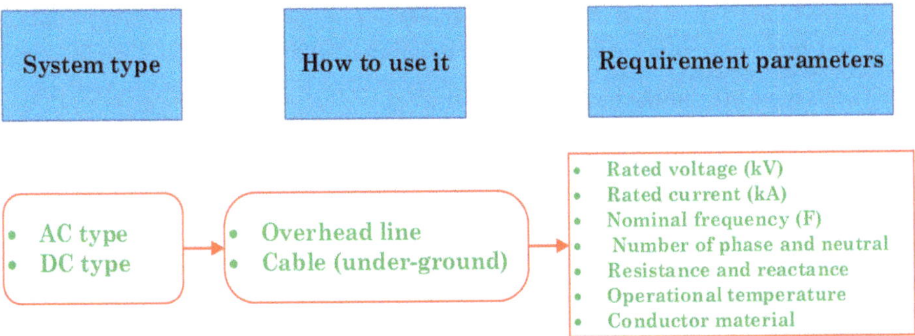

Fig. 2.3 Required parameters for modeling single-line mode

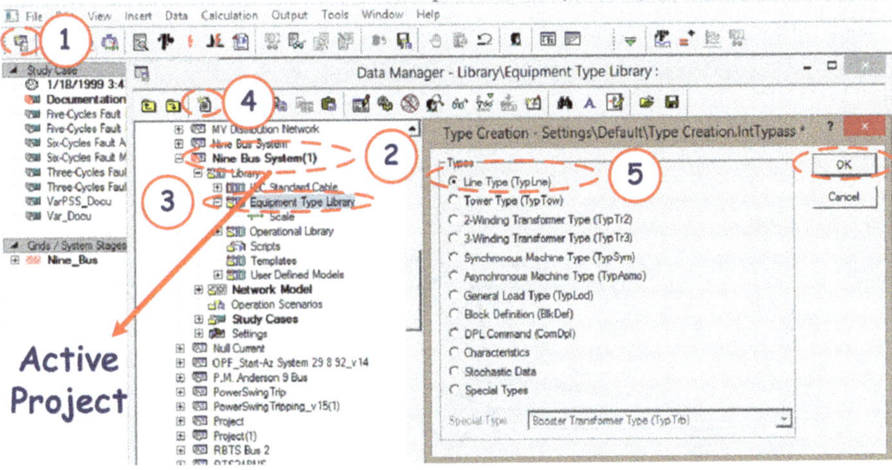

Fig. 2.4 Process of creating global line type

The required steps to modify the consider parameters are as follows:

- *Double-clicking on the desired line*;
- *Go to the "Basic Data" tab and select "New Project Type-Line Type" from "Type" section*;
- *Determine desired parameters from "Basic Data" and "Load Flow" sheet.*

Users can also add commonly used lines to the project library as a global type. To this end, the global type of the power line is created within the active project by taking the following steps.

- *Go to the active project's library from "Data Manager" window* (marked ❶, ❷ in Fig. 2.4);
- *Go to "Equipment Type Library" section then select the "New Object" icon* (marked ❸, ❹ in Fig. 2.4);

- *Select "Line Type" and then Ok* (marked ❺ in Fig. 2.4);
- *Enter the required information on the opening window.*

These steps are shown in Fig. 2.4.

2.2.2.2 Transmission Tower Line System

To analyze and evaluate transmission systems, it is necessary to check the status of power lines on transmission towers. A schematic of the connection of power lines to a transmission tower is shown in Fig. 2.5. As shown in this figure, double circuit transmission lines are connected to each tower. The ability of line coupling technology can be used to study the effects of the double circuit transmission lines in PowerFactory. In addition, the Cartesian coordinate can be used to determine the size of the tower and uses it in the software. Therefore, the tower base is used as the origin of the coordinate system.

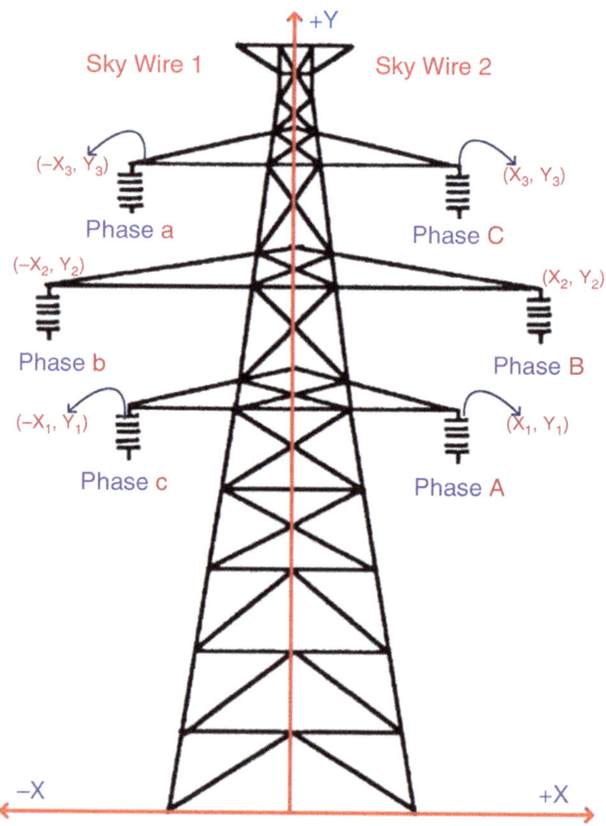

Fig. 2.5 Schematic of the transmission tower

The required steps for analyzing transmission tower line systems are as follows:

- *Select two parallel lines then right-click on the marked lines and press "Line Couplings" from "Define" option* (marked ❶, ❷ in Fig. 2.6);
- *Go to the active project's library from "Data Manager" window then go to the "Equipment Type Library" section* (marked ❸ in Fig. 2.6);
- *Create a tower geometry type using the "New Object" button* (marked ❹, ❺ in Fig. 2.6);
- *Enter the tower information on the opening page, according to* Fig. 2.7;
- *Double-clicking on the created tower type* (marked ❶, ❷ in Fig. 2.8);
- *One of the double circuit transmission lines must be set to A-b-c sequence and the other to C-B-A sequence* (marked ❸, ❹ in Fig. 2.8).

These steps are shown in Figs. 2.6, 2.7, and 2.8.

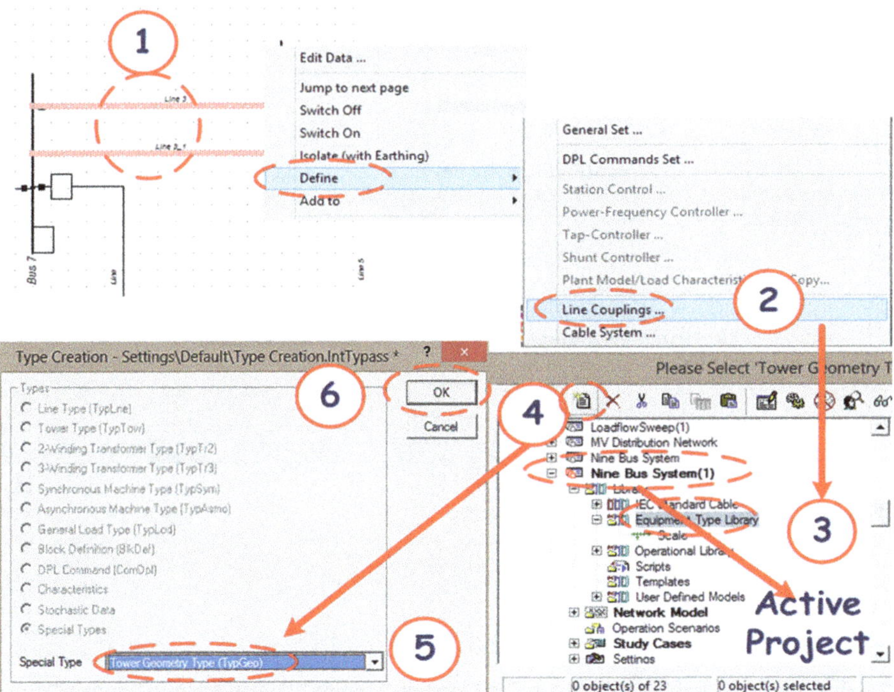

Fig. 2.6 Process of creating transmission tower line system

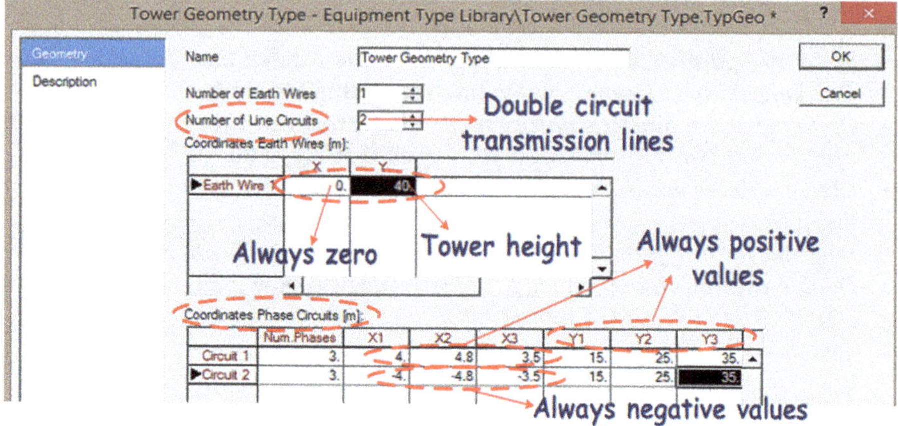

Fig. 2.7 Tips on entering tower information

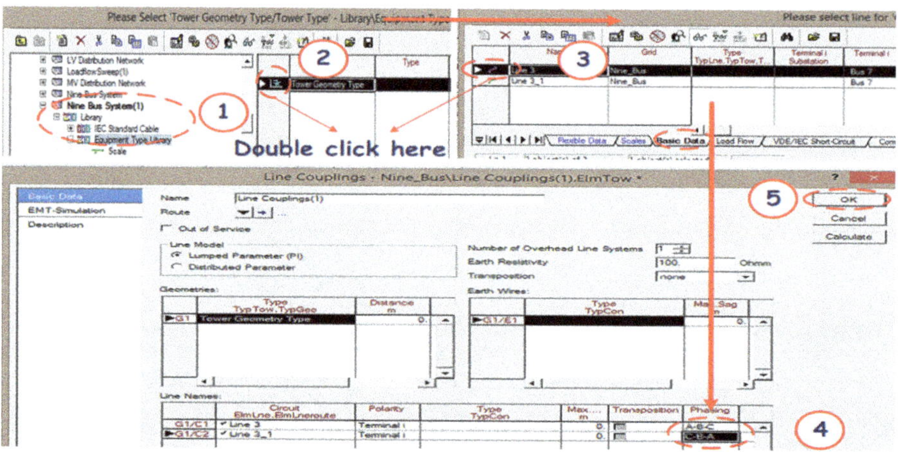

Fig. 2.8 Process of determining lines sequence

2.2.3 *Load Modeling*

Electrical loads are a vital part of the power system. The most important components of these elements are lightbulbs, heaters, resistors, and motors. It is necessary to model the network loads in different modes for the rational evaluation of the power system. These models are shown in Fig. 2.9. The implementation steps of each model are outlined below.

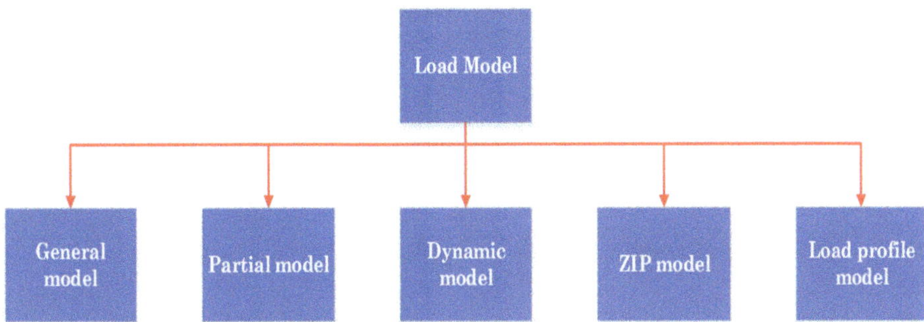

Fig. 2.9 Different load models in power system studies

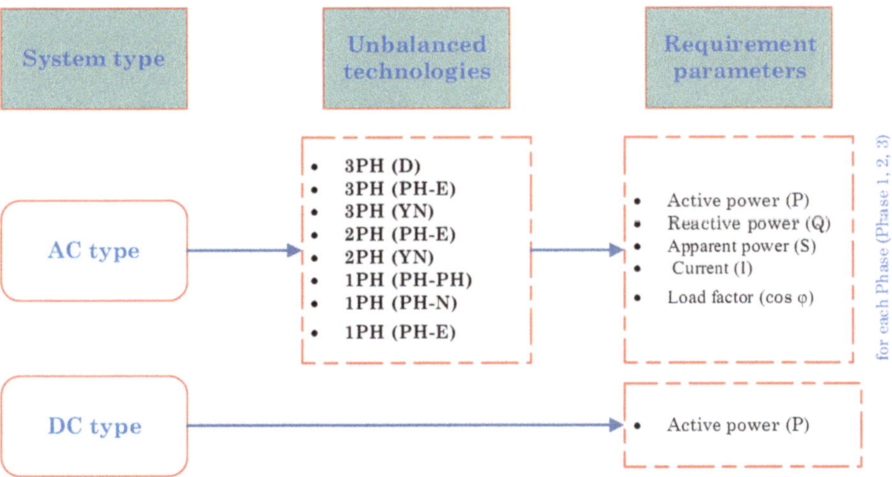

Fig. 2.10 Different type and technologies in general load model

2.2.3.1 General Model

This option is a general way to model balanced and unbalanced loads in power systems. In unbalanced mode, we can use two types of systems as well as eight different technologies to cover the various conditions in the power system. For AC load modeling, two distinct components must be identified from the parameters of active power (P), reactive power (Q), apparent power (S), current (I), and load factor (cosφ). However, for DC load modeling, only the active power is required. These technologies and requirement parameters are given in Fig. 2.10.

The Fig. 2.11 is describing how to utilize a general load type (*TypLod*). The steps of model development are:

- *Double-clicking on the desired load*;
- *Go to the "Basic Data" tab and select "New Project Type-General Load Type" from "Type" section* (marked ❶–❸ in Fig. 2.11);

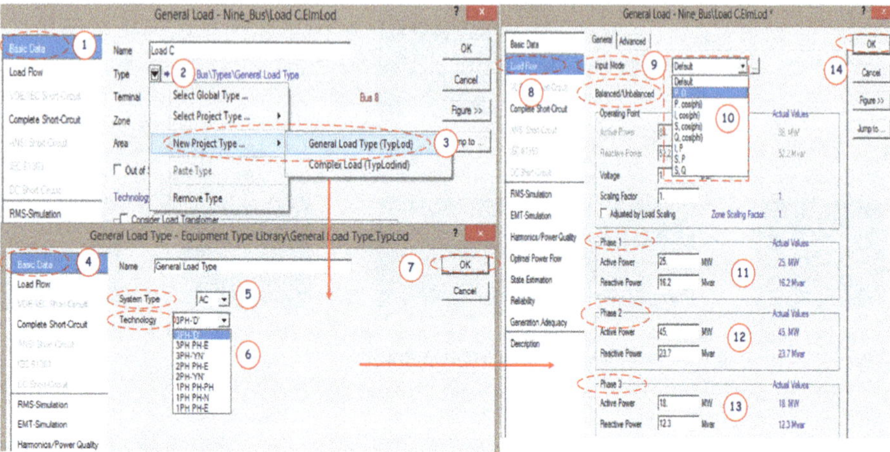

Fig. 2.11 Process of creating general loads

- *Determine "System Type" and "Technology"* (marked ❺, ❻ in Fig. 2.11);
- *Go to the "load flow" option and determine "Balanced/Unbalanced System" as well as technical parameters (at least two values for each phase)* (marked ❾–❶❸ in Fig. 2.11).

2.2.3.2 Partial Model

This model is an evolved method than the general load model for more accurate modeling of low voltage subscribers. Using this option, we can model the exact number of subscribers connected to the low voltage side of the distribution transformer based on the power consumption and location of each subscriber. The partial load concept is shown in Fig. 2.12. As it can be seen in Fig. 2.12, low voltage side of the distribution transformer can host some partial loads. These loads are located at a distance of 10, 40, and 70% from the beginning of the line 1–2, with different power consumption. The required steps to apply fixed partial loads in the power system include:

- *Double-clicking on the target line to connect partial loads*;
- *Go to the "Load Flow" tab and select "Line Loads" box* (marked ❶, ❷ in Fig. 2.13);
- *Select the "New Object" icon and define new partial load* (marked ❸ in Fig. 2.13);
- *In "Partial LV-Load" panel, at first, determine customers position on the line* (marked ❺ in Fig. 2.13).

Fig. 2.12 Partial loads modeling

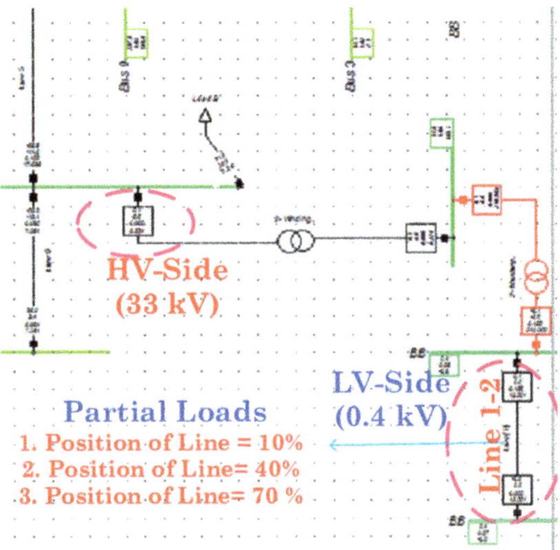

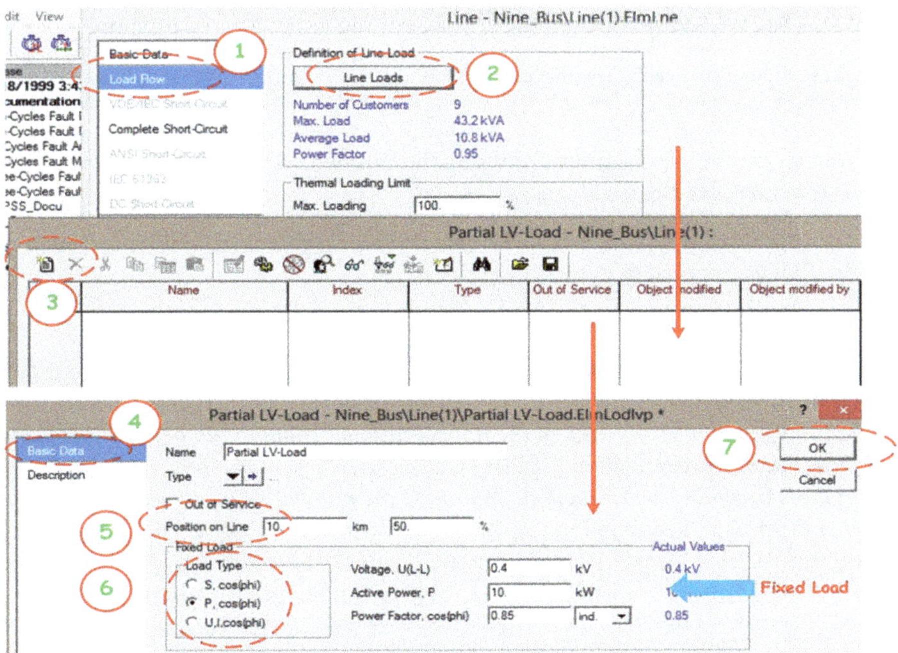

Fig. 2.13 Process of creating fixed partial loads

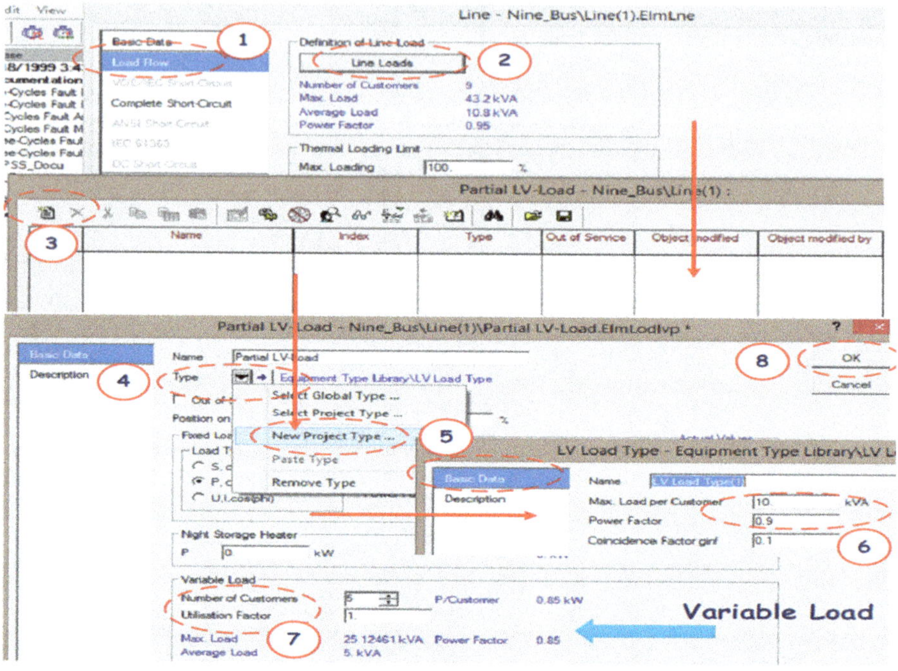

Fig. 2.14 Process of creating variable partial loads

We can determine customers' consumption in two ways: fixed load or variable load. To model fixed partial loads:

- *Active power, apparent power, current, and power factor are used for modeling fixed loads* (marked ❻ in Fig. 2.13).

To model variable partial loads:

- *Go to the "Basic Data" tab and select "New Project Type" from "Type" section* (marked ❹, ❺ in Fig. 2.14);
- *In the page that opens, assign the appropriate maximum load per customer and power factor* (marked ❻ in Fig. 2.14);
- *In the "Variable Load" section, determine the number of customers and utilization factor* (marked ❼ in Fig. 2.14).

These steps are shown in Figs. 2.13 and 2.14.

2.2.3.3 Dynamic Model

The dynamic load model has a significant role in mid-term and long-term transient stability analysis of the power system. On the other hand, dynamic load model is a specific tool to deal with the emerging challenges in the power system. For example, the use of electric heating loads has been increased dramatically over the last decades. Excessive application of these loads can affect voltage stability. Hence, the dynamic model can be useful for analyzing the role of thermal loads in system stability and in general, for examining transient stability of the power system.

The dynamic active and reactive loads model can be defined by three parameters, steady state load-voltage dependence, load-recovery time constant, and transient load-voltage dependence. The mathematical equations of the dynamic load model are presented as a voltage-dependent nonlinear method in Eqs. (2.1a, 2.1b, 2.1c, and 2.1d) as follows [18]:

$$T_p \frac{dP_r}{dt} + P_r = P_0 \left(\frac{U}{U_0} \right)^{\alpha_s} - P_0 \left(\frac{U}{U_0} \right)^{\alpha_t} \tag{2.1a}$$

$$Pl = P_r + P_0 \left(\frac{U}{U_0} \right)^{\alpha_t} \tag{2.1b}$$

$$T_q \frac{dQ_r}{dt} + Q_r = Q_0 \left(\frac{U}{U_0} \right)^{\beta_s} - Q_0 \left(\frac{U}{U_0} \right)^{\beta_t} \tag{2.1c}$$

$$Ql = Q_r + Q_0 \left(\frac{U}{U_0} \right)^{\beta_t} \tag{2.1d}$$

where, P_0 and U_0 are the power consumption and voltage before a voltage deviation (the base values); P_r and Q_r are the active and reactive power recovery; Pl and Ql are the total active and reactive power response; T_p and T_q are the active and reactive load recovery time constant; α_t and $ß_t$ are the transient active and reactive load-voltage dependence; α_s and $ß_s$ are the steady state active and reactive load-voltage dependence. Constants α and ß are variable for different conditions and are calculated by using curve fitting. According to [19], these constants have been estimated for several sample studies on Sweden's power system. The summary of results is presented in Table 2.2.

Table 2.2 Identified constant parameters for active and reactive response under different voltage steps

$\Delta V/V_0$ (%)	T_p (s)	α_t	α_s	T_q (s)	$ß_t$	$ß_s$
−1.8	135	1.36	0.25	9	−181.18	7.9
+1.9	40	1.7	−0.1	256	−687.2	9475.6
−3.7	61	1.31	−0.16	88	87.67	31.86
+3.7	74	1.35	−0.54	105	104.5	−148.56
−5.3	70	1.65	−0.32	78	77.89	19.35
+5.4	78	1.6	−0.08	94	94.18	−49.75

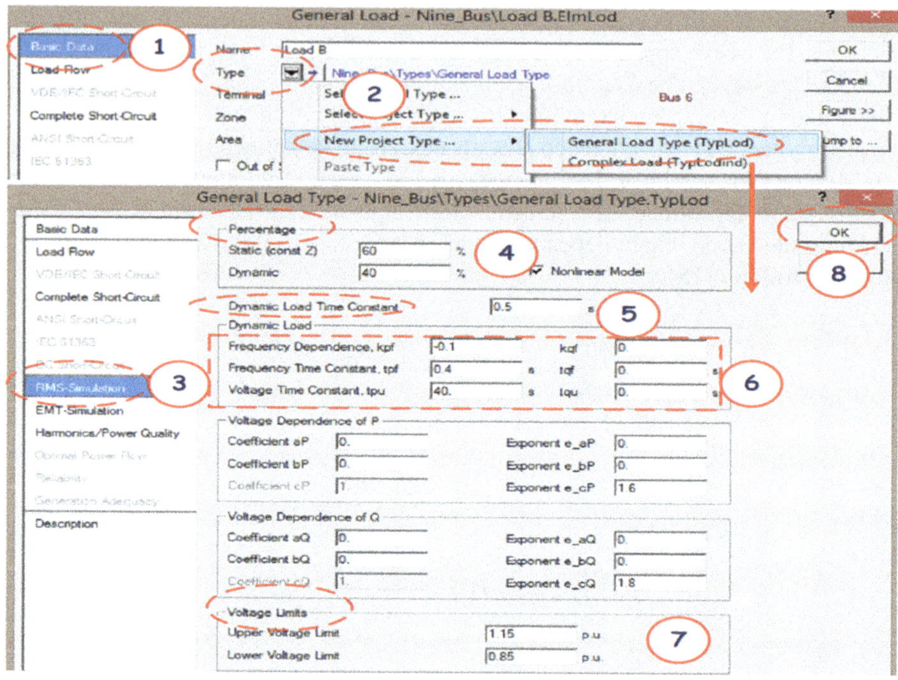

Fig. 2.15 Process of creating dynamic loads

The required steps to apply dynamic loads in the power system include:

- *Double-clicking on the desired load*;
- *Go to the "Basic Data" tab and select "New Project Type-General Load Type" from "Type" section* (marked ❶, ❷ in Fig. 2.15);
- *Go to the "RMS-Simulation" tab and determine the percentage of the dynamic load and static load* (marked ❸, ❹ in Fig. 2.15);
- *In the "Dynamic Load" section, determine the time constant, and the coefficients for the time constant of voltage and frequency. These coefficients are determined by the type of load and the conditions of the power system, which are presented in* Table 2.2 *for a number of sampling points in Sweden's power system* (marked ❺, ❻ in Fig. 2.15);
- *This load model is only valid within the specified voltage range. We can change the desired range by "Voltage Limits" section* (marked ❼ in Fig. 2.15).

These steps are shown in Fig. 2.15. The behavior of the dynamic loads under extended voltage ranges is shown in Fig. 2.16. This figure can be described by the nonlinear equations presented in Eqs. (2.2a, 2.2b, 2.2c, 2.2d, and 2.2e).

$$P = k.P_{\text{out}}; \quad Q = k.Q_{\text{out}} \tag{2.2a}$$

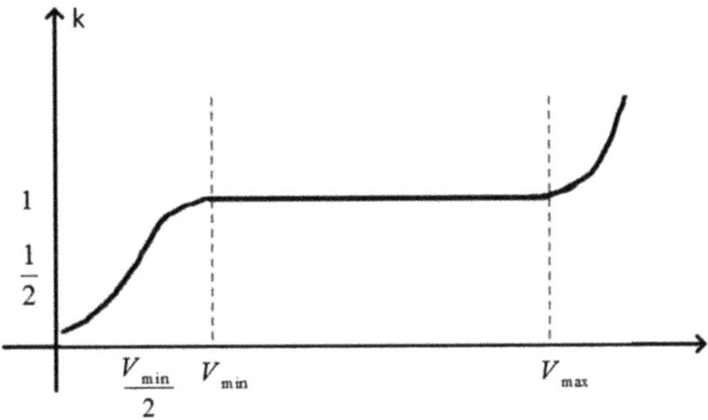

Fig. 2.16 Voltage approximations used in dynamic load model

$$k = 1; \quad V_{min} \le V \le V_{max} \tag{2.2b}$$

$$k = \frac{2\left|V\right|^2}{V_{min}^2}; \quad 0 \le V \le \frac{V_{min}}{2} \tag{2.2c}$$

$$k = 1 - 2\left(\frac{\left|V\right| \;\; V_{min}}{V_{min}}\right)^2; \quad \frac{V_{min}}{2} \le V \le V_{min} \tag{2.2d}$$

$$k = 1 + \left(V - V_{max}\right)^2; \quad V \ge V_{max} \tag{2.2e}$$

2.2.3.4 ZIP Model

The ZIP model is the best option for analyzing static loads. We can investigate three common modes of loads in the power system, namely constant impedance (Z), constant current (I), and constant power (P) by applying this model. The ZIP load model is formulated in Eqs. (2.3a and 2.3b) as follows [13]:

$$P = P_0\left(aP\left(\frac{V}{V_0}\right)^{e\_aP} + bP\left(\frac{V}{V_0}\right)^{e\_bP} + \left(1 - aP - bP\right)\left(\frac{V}{V_0}\right)^{e\_cP}\right) \tag{2.3a}$$

$$Q = Q_0\left(aQ\left(\frac{V}{V_0}\right)^{e\_aQ} + bQ\left(\frac{V}{V_0}\right)^{e\_bQ} + \left(1 - aQ - bQ\right)\left(\frac{V}{V_0}\right)^{e\_cQ}\right) \tag{2.3b}$$

where $1 - aP - bP = cP; \quad 1 - aQ - bQ = cQ$

Table 2.3 Typical values for exponents *e_cP* and *e_cQ*, for different load components

Load component	*e_cP*	*e_cQ*
Air conditioner	0.5	2.5
Resistance space heater	2	0
Fluorescent lighting	1	3
Pumps, fans other motors	0.08	1.6
Large industrial motors	0.05	0.5
Small industrial motors	0.1	0.6

In the sabove equation, *P* and *Q* are the active and reactive powers at rated voltage *V*; P_0 and Q_0 are the active and reactive powers at operating voltage V_0; *aP*, *bP*, and *cP* are the ZIP coefficients for active power; *aQ*, *bQ*, and *cQ* are the ZIP coefficients for reactive power. The behavior of various loads is determined by changing the respective exponents (*e_aP*, *e_bP*, *e_cP*; *e_aQ*, *e_bQ*, *e_cQ*).

For modeling power system loads in DIgSILENT PowerFactory, the coefficients of *aP*, *bP*, *aQ* and *bQ* are assumed to be zero, *cP* and *cQ* coefficient are assumed to be one. Therefore, the determination of the load characteristics depends only on the exponential coefficients (*e_cP* and *e_cQ*). In Eqs. (2.3a and 2.3b), for modeling constant power loads, exponential coefficients *e_cP* and *e_cQ* must be equal to zero. For modeling of constant current loads, these coefficients must be equal to one and for modeling of constant impedance loads, these coefficients must be equal to two. Based on the previous researches, usual values for the exponential coefficients *e_cP* and *e_cQ* for different load components are given in Table 2.3 [20].

All provided descriptions can be applied in DIgSILENT PowerFactory, according to Fig. 2.17.

- *Double-clicking on the desired load*;
- *Go to the "Basic Data" tab and select "New Project Type-General Load Type" from "Type" section* (marked ❶, ❷ in Fig. 2.17);
- *Go to the "Load Flow" tab and determine the respective exponents* (marked ❸–❺ in Fig. 2.17).

2.2.3.5 Load Profile Model

Power systems operator should monitor the status of the network every hour and prevent unintentional technical errors in the network. The best way to evaluate power consumption in the network is to use the 24-h load profile. For example, in transmission networks, due to the change of active and reactive loads within 24 h, shunt reactor and shunt capacitor should be used to maintain the voltage in the acceptable range. Hence, the use of load curves would be useful in conducting technical analysis. To allocate a 24-h active load profile for the desired load, the following steps should be performed:

- *Double-clicking on the desired load*;
- *Go to the "Operating Point" section in the "Load Flow" tab* (marked ❶ in Fig. 2.18);

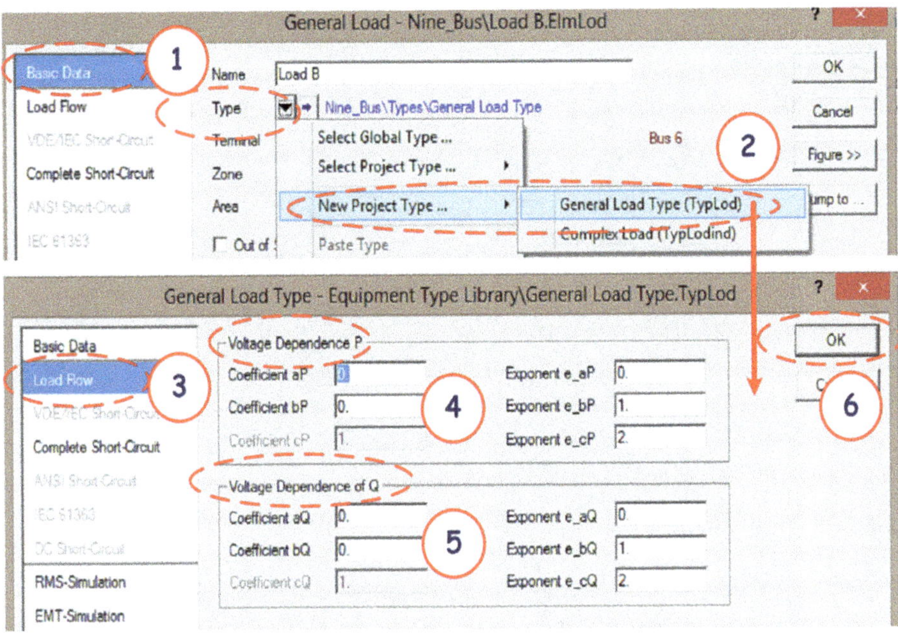

Fig. 2.17 Process of creating static loads (ZIP model)

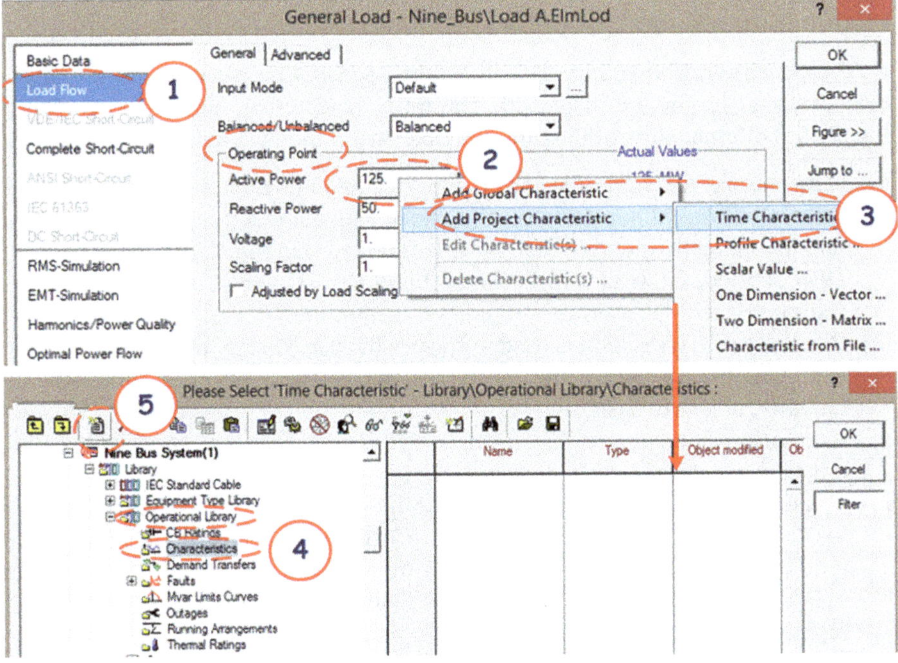

Fig. 2.18 Process of creating load profile

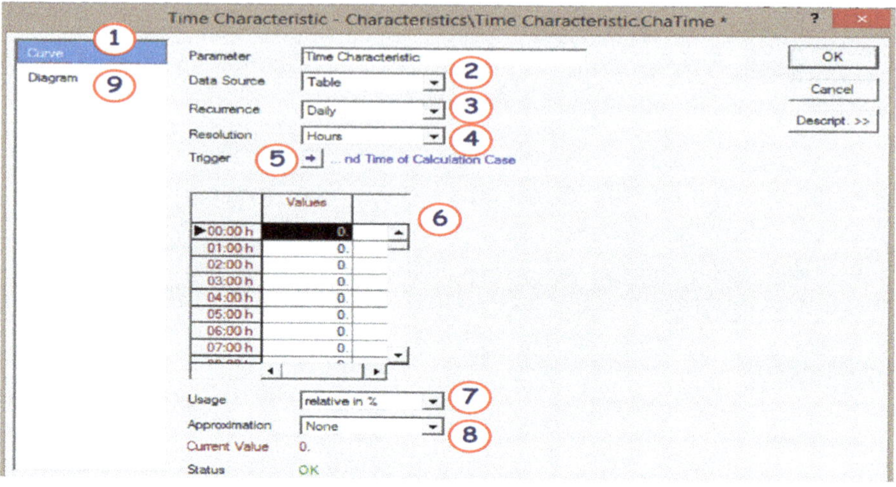

Fig. 2.19 Time characteristic window

- *Right-click on the "Active Power" box then select "Add Project Characteristic-Time Characteristics"* (marked ❷, ❸ in Fig. 2.18);
- *The data manager window opens by selecting "Time Characteristics" option*;
- *In the data manager window, go to the "Operational Library" and select "New Object" button in the "Characteristics" folder* (marked ❹, ❺ in Fig. 2.18);
- *After the above steps, a new object from "ChaTime" class opens automatically.*

These steps are shown in Fig. 2.18. In the opened window, required settings to enter the load profile are displayed. The time characteristic window is shown in Fig. 2.19. Different parts of the time characteristic window are described below.

(a) **Data source**: A load profile can be imported into PowerFactory in two different ways. By using "*Table Data*" option, information about the load profile can be imported directly into the existing table (marked ❻ in Fig. 2.19). By using "*File Data*" option, the load profile can be imported into PowerFactory through an external Excel file (∗.csv).

(b) **Recurrence**: This option is used to specify the periodically recurring characteristic. The periodically recurring values can be specified on a daily, weekly, monthly, or yearly basis.

(c) **Resolution**: If the repeat "*Recurrence*" is set to daily values, the load information can be entered hourly or minutes.

(d) **Usage**: This option specifies how the imported values will affect the desired load characteristic. If this option is set on "*Relative,*" the imported values in the time characteristic table are multiplied by the specified parameter. But if this option is set on "*Absolute,*" the imported values in the time characteristic table replace the desired parameter.

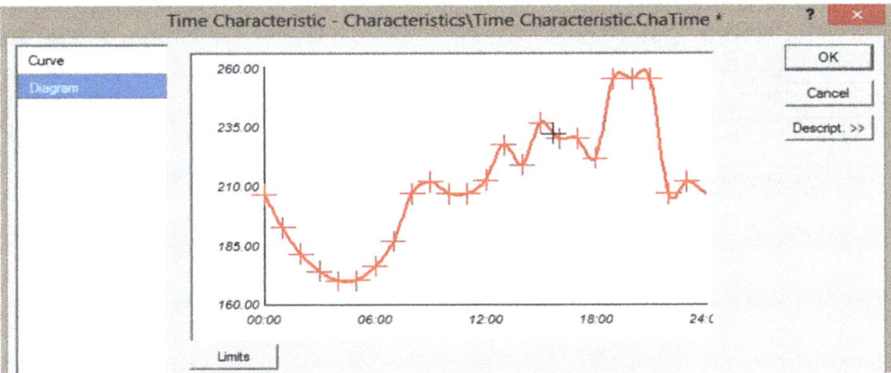

Fig. 2.20 Daily active load curve in PowerFactory

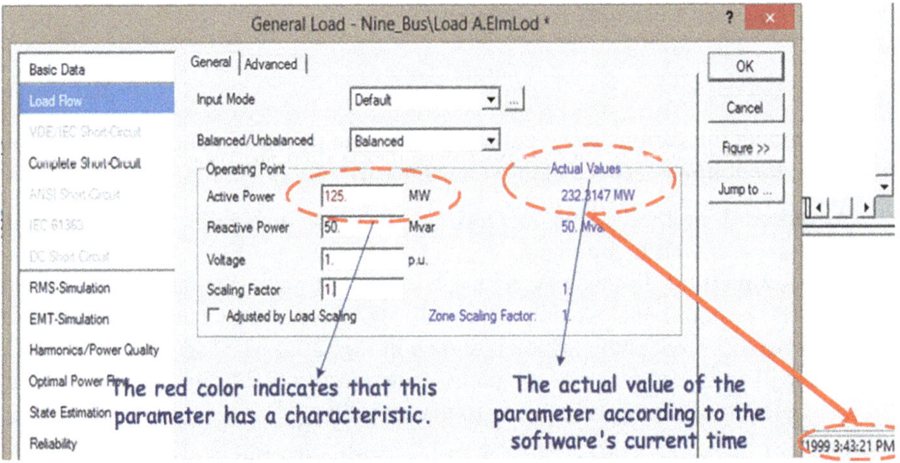

Fig. 2.21 Changes made in the load information sheet

(e) **Approximation**: This option specifies the rational approximation method for the intermediate intervals. Available approximation methods include linear, polynomial, spline, and Hermite.

For example, the daily active load curve of a sample distribution system that imported into PowerFactory through the described steps is shown in Fig. 2.20. The load characteristic is created after importing all information and confirming the time characteristic window.

After performing all the above steps, the desired load information sheet will be displayed. As shown in Fig. 2.21, the color of the desired parameter (active power) is turned red. In addition, the actual value of the 24-h load profile is shown with blue color according to the software time.

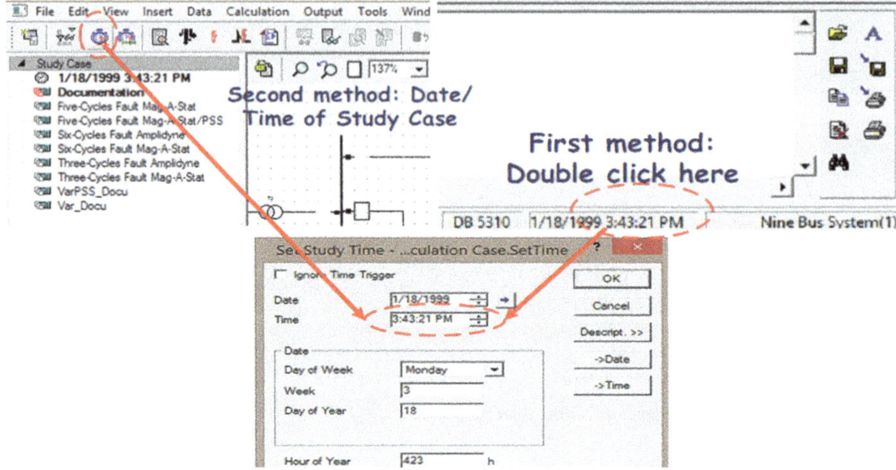

Fig. 2.22 The methods of setting the time manually

To conduct the various technical analysis on the power system at different time intervals, the following two methods can be used:

I. **Automated method**: This method can be done using the *Quasi Dynamic Analysis* or the DPL command.
II. **Manual method**: In this way, first, the software time is changed manually, and then the required analysis is performed. The system time can be manually set from two different paths, which is shown in Fig. 2.22. The "*set study time*" window will open using any of the paths shown in Fig. 2.22. Finally, the system's active load will change according to the set time.

To make the desired changes on the load profile, the following actions need to be taken:

- *Double-clicking on the desired load*;
- *Go to the "Operating Point" section in the "Load Flow" tab* (marked ❶ in Fig. 2.23);
- *Right-click on the "Active Power" box then select "Edit Characteristics"* (marked ❷, ❸ in Fig. 2.23);
- *In the opened box, double-clicking on the time characteristic* (marked ❹ in Fig. 2.23).

These steps are shown in Fig. 2.23.

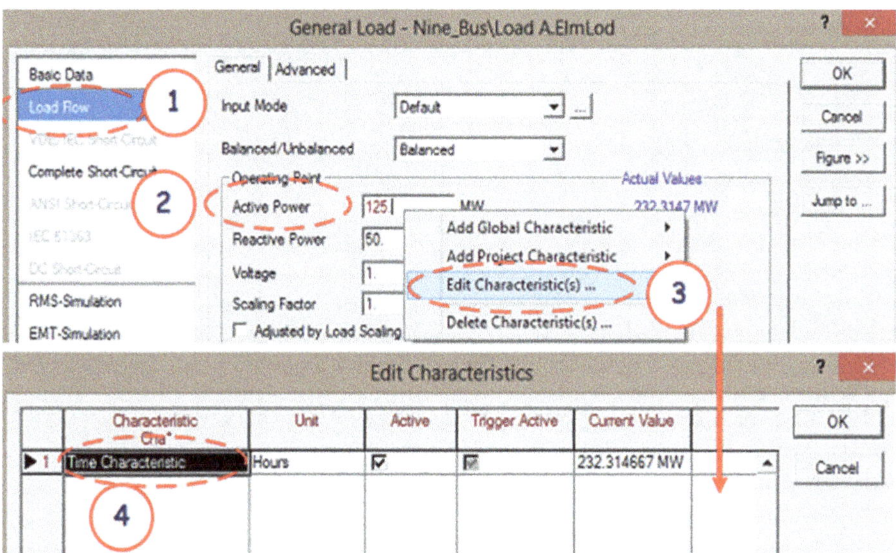

Fig. 2.23 Process of editing load profile

2.3 Useful Analysis in Power Systems

In this section, we will investigate some of the important tools related to the power system's loads and lines.

2.3.1 Time Sweep Analysis

Time sweep is one of the most applicable methods for analyzing the behavior of low-voltage (LV) distribution network's feeders during 24-h. During this analysis, the behavior of LV distribution network's feeders is evaluated from the perspective of the upstream substations. As mentioned in the previous section, each consumer connected to the network's feeders has a separate time characteristic. In this section, we want to evaluate the total active load and losses status of each feeder according to the different time characteristics of each consumer within 24-h from the network's operator perspective. To this end, the following steps should be performed:

- *Run a sample distribution network with several feeders and different time characteristics in PowerFactory. We can use the sample network available in DIgSILENT*;
- *Press the "Edit Relevant Objects for Calculation" button located on the main menu and select "Feeder" icon* (marked ❶ in Fig. 2.24a);

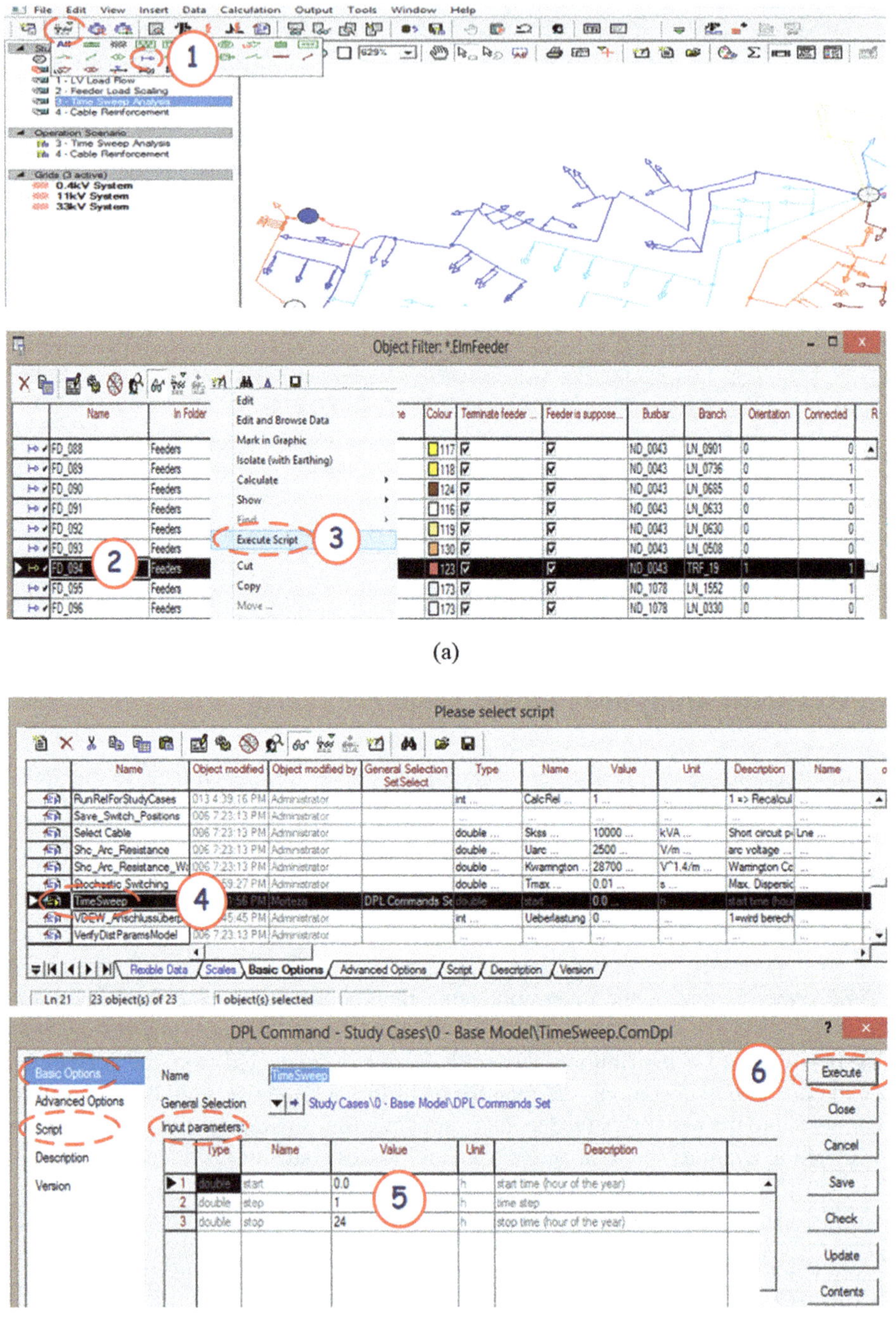

Fig. 2.24 Process of time sweep analysis (**a**) access to the command and (**b**) required information

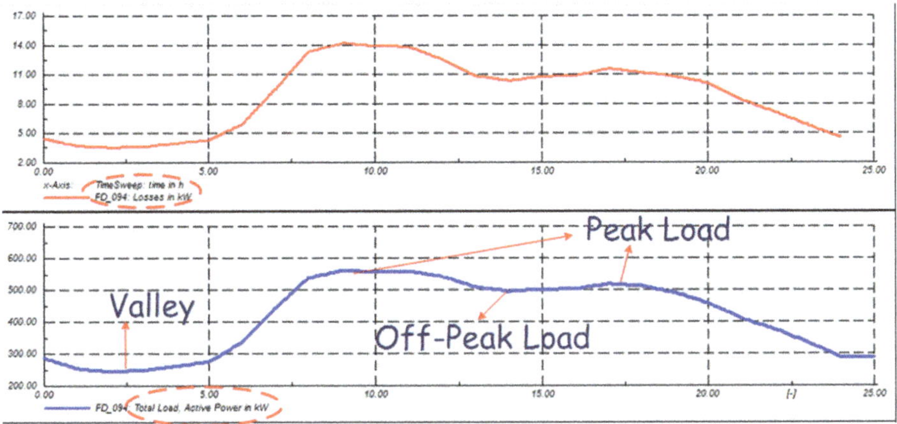

Fig. 2.25 Total active power and losses graph

- *By executing step 2, the list of all feeders is displayed. Select the desired feeder from the provided list*;
- *Right-click on the selected feeder then choose "Execute Script" option* (marked ❷, ❸ in Fig. 2.24a);
- *Select "Time Sweep" from the provided scripts* (marked ❹ in Fig. 2.24b);
- *Through the "Basic Options," input parameters including start and stop times of the simulation as well as the simulation steps can be changed* (marked ❺ in Fig. 2.24b);
- *DPL code for Time Sweep analysis is also visible through the "Script" option*;
- *After executing the time sweep analysis, diagrams related to the "Total Active Power" and also "Losses" of the selected feeder are displayed within 24-h.*

Generally, *Time Sweep script* executes *Load Flow Analysis* 25 times from 0 to 24 and measures the amount of active power flowing through the feeder at each time. Finally, the results of the *Load Flow Analysis* are displayed in a graph. These steps are shown in Figs. 2.24 and 2.25.

2.3.2 Cable Sizing Analysis

The *Cable Sizing Analysis* is applied to verify the type of the existed cables or to determine the optimum cables in different parts of the power system with the aim of reducing line loading and improving the voltage profile of feeders. The "LV Distribution Network," one of the PowerFactory examples, is used to demonstrate how this analysis is applied.

At first, to check the network status from the view of line loading, follow these steps:

- *Click the "Diagram Colouring" icon located on the main menu. The diagram colouring dialog will appear;*
- *Select "Load Flow" tab and then choose "Results" by means of the first drop-down list on the "Other" box;*
- *Select "Voltages/Loading" from the second drop-down list on the "Other" box;*
- *Click OK to save your changes and close the diagram colouring dialogue;*
- *Finally, run the load flow analysis command.*

It can be seen that the feeders 231, 158, and 239 are not in good condition with maximum loading of 111.74, 118.04, and 138.45%, respectively. Hence, *Cable Sizing Analysis* should be carried out for these feeders. In the following, for example, *Cable Sizing Analysis* is performed for feeder 158.

To perform the *Cable Sizing Analysis*, press the "*Calculation*" button located on the main menu, and select "*Cable Sizing*" command. Then the cable sizing dialogue will pop-up, which contains the settings for Basic Options, Constraints, Output, and Advanced Options. The required settings to run *Cable Sizing Analysis* correctly are as follows:

I. **"Basic options" tab**

Method section: *Cable Sizing Analysis* can be done by two methods of "*International Standards*" and "*Cable Reinforcement*". Both methods verify the suitability of the assigned line types or recommend the new line types to the user. However, in the international standards method network constraints such as voltage and thermal constraints are determined according to the selected standards, whereas, in the cable reinforcement method these constraints are determined by the user. One of the available methods must be chosen to perform this analysis. By choosing the "Cable Reinforcement" option, the user would have more freedom to determine the network conditions. Therefore, we recommend using this option for network analysis.

Feeders section: After choosing the "*Cable Reinforcement*" option, select feeder 158 using the "*Selected Feeders*" option.

Mode section: By using this section, PowerFactory can confirm the suitability of the existing line types and also can suggest suitable lines instead of existing lines. "*Verification*" option only evaluates the suitability of the existing line types. On the contrary, by selecting "*Recommendation*" option, PowerFactory will automatically choose the suitable line types with a voltage rating suitable for the line element. Some sample line types are available in the global library. We can either use these line types or add suggested line types to "*IEC Standard Cable*" folder.

Network representation section: Using this box, the overall state of the network, such as balanced or unbalanced, and the required conditions for *Load Flow* and *Short-Circuit Analysis* must be determined.

II. **"Constraints" tab**

In this section, the maximum allowable thermal loading of lines and the voltage drop of feeders must be determined. It is best to apply all constraints using the "*Global Limit*" mode. Typically, the maximum thermal loading for all lines is equal to 80%, the lower voltage limit for all terminals is equal to 0.95 pu, and the maximum voltage drop for the selected feeders is equal to 5%.

III. **"Output" tab**

If "*Report Only*" option is selected, only suggested line types are listed in the output window. Nevertheless, if the next two options are selected, optimal modifications will be made to the system topology and report will be presented based on the changes.

IV. **"Advanced options" tab**

Network consistency section: In consistency check option 1 (sum of feeding cables ≥ sum of leaving cables), the sum of the cross-sectional area (or nominal current) of the input cables to each terminal shall be greater than the sum of the cross-sectional area (or nominal current) of the output cables from that terminal. For consistency check option 2 (smallest feeding cables ≥ biggest leaving cables), the smallest input cable to a terminal must be larger than the largest output cable from that terminal.

It can be determined whether the analysis is based on the cross-section or the nominal current of the lines using the "*Criteria*" box. Usually, in many studies, the cross-section option is selected.

2.3.2.1 Cable Sizing Analysis Outputs

As shown in Fig. 2.26, the results of the *Cable Sizing Analysis* are presented in the output window. For feeder 158 (FD_158), the list of optimal cable types to replace existing types is presented. Using the suggested cable types by the software, the maximum loading of FD_158 can be reduced from 118.04 to 42.79%.

Project: LV Distribution Network Feeders : FD_158 Annex: / 4

Recommendation :

Name	Existing Cable Type	Recommended Cable Parameters Type	Rated Current [kA]	Rated Voltage [kV]	Cross Section [mm^2]	Calculation Max. Loading [%]	Cost [$]
LN_1320						42.79	
1:	NAYY 1x70RM 0.6/1kV ir	NKY 1x300rm 0.6/1kV it	0.57	1.00	300.00		0.00
2:	NAYY 1x70RM 0.6/1kV ir	NYY 1x300RM 0.6/1kV ir	0.63	1.00	300.00		0.25
LN_0140						39.35	
1:	NAYY 1x70RM 0.6/1kV ir	NAKY 1x300rm 0.6/1kV it	0.45	1.00	300.00		0.00
LN_1291	NK NAYCWY SE/150 3x150 Al	NYY 1x300RM 0.6/1kV ir	0.63	1.00	300.00	38.15	0.90
LN_1257	NK NAYCWY SE/150 3x150 Al	NYY 1x300RM 0.6/1kV ir	0.63	1.00	300.00	37.35	0.43
LN_1223	NK NAYCWY SE/150 3x150 Al	NYY 1x300RM 0.6/1kV ir	0.63	1.00	300.00	36.71	1.14
LN_1193	NK NAYCWY SE/150 3x150 Al	NYY 1x300RM 0.6/1kV ir	0.63	1.00	300.00	35.73	1.09
LN_1152	NK NAYCWY SE/150 3x150 Al	NYY 1x300RM 0.6/1kV ir	0.63	1.00	300.00	35.18	0.07
LN_0371						33.25	
1:	NAYY 1x70RM 0.6/1kV ir	NKY 1x500rm 0.6/1kV it	0.73	1.00	500.00		0.00
1:	NAYY 1x70RM 0.6/1kV ir	NYY 1x300RM 0.6/1kV ir	0.63	1.00	300.00		0.24
2:	NAYY 1x70RM 0.6/1kV ir	NYY 1x300RM 0.6/1kV ir	0.63	1.00	300.00		0.72
3:	NAYY 1x70RM 0.6/1kV ir	NYY 1x300RM 0.6/1kV ir	0.63	1.00	300.00		0.81
4:	NAYY 1x70RM 0.6/1kV ir	NYY 1x300RM 0.6/1kV ir	0.63	1.00	300.00		0.25
LN_0305	NK NAYCWY SE/150 3x150 Al	NYY 1x300RM 0.6/1kV ir	0.63	1.00	300.00	31.09	0.98
LN_0129						28.89	
1:	NAYY 1x70RM 0.6/1kV ir	NYY 1x300RM 0.6/1kV ir	0.63	1.00	300.00		0.06
LN_0139						27.68	
1:	NAYY 1x70RM 0.6/1kV ir	NYY 1x300RM 0.6/1kV ir	0.63	1.00	300.00		0.11
LN_0138						27.08	
1:	NAYY 1x70RM 0.6/1kV ir	NYY 1x300RM 0.6/1kV ir	0.63	1.00	300.00		0.23
LN_0137						25.88	

Fig. 2.26 Cable sizing analysis outputs

2.4 Linking DIgSILENT and Excel

2.4.1 Writing to Excel

It will take a long time to manually change the data of loads, generators, and lines in the wide-area power networks. For this reason, creating an interface between Excel and PowerFactory is very useful to change or update the data of different elements. For example, extracting the active and reactive power of network loads in the form of an Excel file is straightforward as follows:

- *Choose one of the network loads and then select "Edit and Browse Data" icon* (marked ❶ in Fig. 2.27a);
- *Select all loads through the "Basic Data" tab and then copy them* (marked ❷, ❸ in Fig. 2.27a);
- *Go to the "Description" tab and paste the copied information into the "Foreign key" column* (marked ❹–❻ in Fig. 2.27a);
- *Press "Data Manager" icon then create a new folder by right-clicking on the admin icon* (marked ❶–❸ in Fig. 2.27b);
- *Select created folder then press "New Objective" icon*;

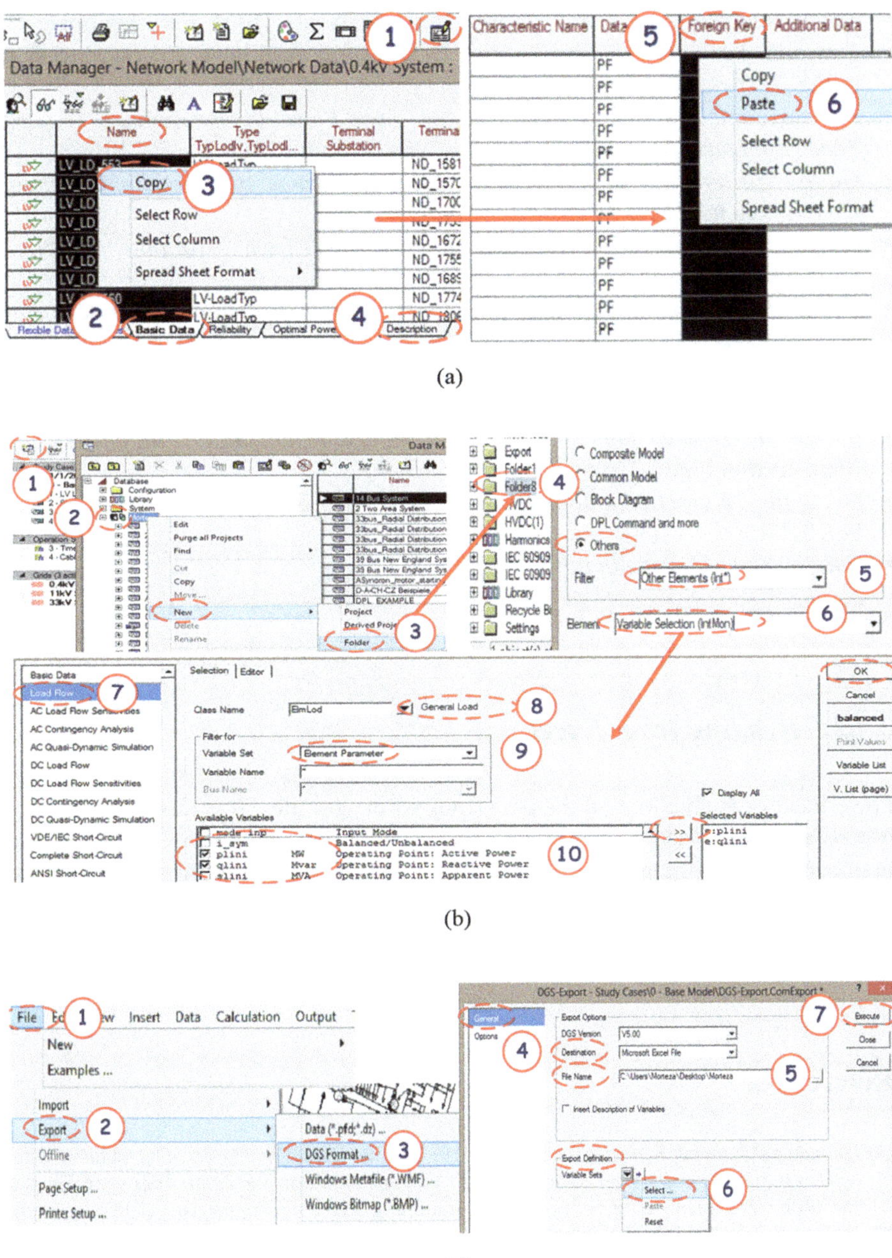

Fig. 2.27 Process of transfer information from PowerFactory to Excel (**a**) defining desired elements, (**b**) determining required data, and (**c**) extracting data

- *Select "Others" option and choose "Other Elements" from first drop-down list then opt "Variable Selection" option from second drop-down list* (marked ❺–❻ in Fig. 2.27b);
- *In the variable selection window, go to the "Load Flow" tab then choose "General Load (ElmLod)" from the "Class Name" menu* (marked ❽ in Fig. 2.27b);
- *Go to "Variable Set" box the select "Element Parameter" from the drop-down list* (marked ❾ in Fig. 2.27b);
- *Choose the desired parameter, such as active power from "Available Variables" box* (marked ❶⓿ in Fig. 2.27b);
- *Select "DGS Format" option from "Export" section located on the "File" menu* (marked ❶–❸ in Fig. 2.27c);
- *In the opened window, choose "Microsoft Excel File" option from "Destination" box* (marked ❹ in Fig. 2.27c);
- *Specify the desired path and name for the excel file from "File Name" box* (marked ❺ in Fig. 2.27c);
- *Go to the "Export Definition" section and select created folder* (marked ❻ in Fig. 2.27c);
- *Finally, press Execute button.*

These steps are shown in Fig. 2.27a–c.

2.4.2 Reading from Excel

PowerFactory is able to read the data from *xls* or *xlsx* files. It is also convenient to use *xls* as an interface between PowerFactory and other platforms. To this end, at first, according to the presented steps in the previous sub-section, create an Excel file with the considered parameters, and then the desired changes must be made to that file. Finally, upload the desired Excel file via the "*Import*" command located on the "*File*" section.

References

1. Chiradeja P, Yoomak S, Ngaopitakkul A (2017) Optimal allocation of multi-DG on distribution system reliability and power losses using differential evolution algorithm. Paper presented at the 4th International Conference on Power and Energy Systems Engineering (CPESE), Berlin
2. Roldan-Fernandez JM, Gonzalez-Longatt FM, Rueda JL et al (2014) Modelling of transmission systems under unsymmetrical conditions and contingency analysis using DIgSILENT PowerFactory. In: PowerFactory Applications for Power System Analysis. pp 27–52
3. Chathurangi D, Jayatunga U, Rathnayake M et al (2018) Potential power quality impacts on LV distribution networks with high penetration levels of solar PV. Paper presented at the 18th International Conference on Harmonics and Quality of Power (ICHQP), Ljubljana

4. Dillingham C, Silva-Saravia H, Pulgar-Painemal H (2017) Determining wide-area signals and locations of regulating devices to damp inter-area oscillations through eigenvalue sensitivity analysis using digsilent programming language. In: Advanced Smart Grid Functionalities Based on PowerFactory. pp 153–179
5. Teimourzadeh S, Mohammadi-Ivatloo B (2014) Probabilistic power flow module for PowerFactory DIgSILENT. In: PowerFactory Applications for Power System Analysis. pp 61–84
6. Jannesar MR, Sedighi A, Savaghebi M et al (2019) Optimal probabilistic planning of passive harmonic filters in distribution networks with high penetration of photovoltaic generation. Int J Electr Power Energy Syst 110:332–348
7. Kim I, Regassa R, Harley RG (2015) The modeling of distribution feeders enhanced by distributed generation in DIgSILENT. Paper presented at the 42nd Photovoltaic Specialist Conference (PVSC), New Orleans
8. Velaga YN, Chen A, Sen PK et al (2018) Transmission-distribution co-simulation: model validation with standalone simulation. Paper presented at the North American Power Symposium (NAPS)
9. Sørensen PE, Hansen AD, Thomsen K et al (2005) Operation and control of large wind turbines and wind farms. Risoe-R, No. 1532
10. Khajeh KG, Bashar E, Rad AM et al (2017) Integrated model considering effects of zero injection buses and conventional measurements on optimal PMU placement. IEEE Trans Smart Grid 8(2):1006–1013
11. Gonzalez-Longatt F, Rueda JL (2014) PowerFactory applications for power system analysis. Springer, Chem
12. Shotorbani AM, Madadi S, Mohammadi-Ivatloo B (2018) Wide-area measurement, monitoring and control: PMU-based distributed wide-area damping control design based on heuristic optimisation using DIgSILENT PowerFactory. In: Advanced Smart Grid Functionalities Based on PowerFactory. pp 211–240
13. Moreira LT, Mota AA (2004) Load modeling at electric power distribution substations using dynamic load parameters estimation. Int J Electr Power Energy Syst 26(10):805–811
14. Kundur P, Paserba J, Ajjarapu V (2004) Definition and classification of power system stability IEEE/CIGRE joint task force on stability terms and definitions. IEEE Trans Power Syst 19(3):1387–1401
15. Savulescu SC (2009) Real-time stability assessment in modern power system control centers. Wiley-IEEE Press, New York
16. Cepeda J, Rueda J, Colomé G et al (2014) Real-time transient stability assessment based on Centre-of-inertia estimation from PMU measurements. IET Gener Transm Distrib 8(8):1363–1376
17. DIgSILENT GmbH (2014) DIgSILENT PowerFactory Version 15 User's Manual. Gomaringen
18. Karlsson D, Hill DJ (1994) Modelling and identification of nonlinear dynamic loads in power systems. IEEE Trans Power Syst 9(1):157–166
19. Choi B-K, Chiang H-D, Li Y et al (2006) Measurement-based dynamic load models: derivation, comparison, and validation. IEEE Trans Power Syst 21(3):1276–1283
20. Bokhari A, Alkan A, Dogan R et al (2014) Experimental determination of the ZIP coefficients for modern residential, commercial, and industrial loads. IEEE Trans Power Del 29(3):1372–1381

Chapter 3
Modeling and Optimal Operation of Renewable Energy Sources in DIgSILENT PoweFactory

This chapter presents the instruction on how to model renewable energy sources (RES) within the framework of power systems. Different stochastic models will be discussed, such as Weibull distribution function, time series, and generation adequacy in DIgSILENT PowerFactory. The modeling of RES is performed based on some predictions, including ambient temperature, solar radiation, and wind speed.

3.1 Renewable Sources Model in PowerFactory

According to statistics released by the International Energy Agency (IEA), the natural gas and oil are important primary energy carriers of the energy markets. There is also a direct relationship between energy consumption and economic prosperity in each country. To prevent environmental degradation, the IEA has obliged all countries to supply much of their energy needs from renewable energy sources (RES) by 2030. Over the last two decades, the penetration of RES has increased in the power systems due to environmental concerns, nuclear power plant problems, and energy security risks. In line with the energy roadmap outlined by the IEA for a greener future, many countries have set their targets based on the maximum utilization of RES and the gradual elimination of traditional power plants. According to the published results by the International Renewable Energy Agency (IRENA), renewable production capacity has increased in the whole world by about 150 GW since 2001. The additions of renewable power capacity around the world in recent years is shown in Fig. 3.1 [1]. As can be seen from Fig. 3.1, among the renewable technologies, solar and wind resources are of particular interest. The widespread use of RES has created some challenges for the electricity grid, in addition to bringing considerable benefits.

Because of the transmission line's capacity, network structure, and operational constraints, a large amount of renewable production is curtailed, mainly from solar

M. Zare Oskouei, B. Mohammadi-Ivatloo, *Integration of Renewable Energy Sources Into the Power Grid Through PowerFactory*, Power Systems,
https://doi.org/10.1007/978-3-030-44376-4_3

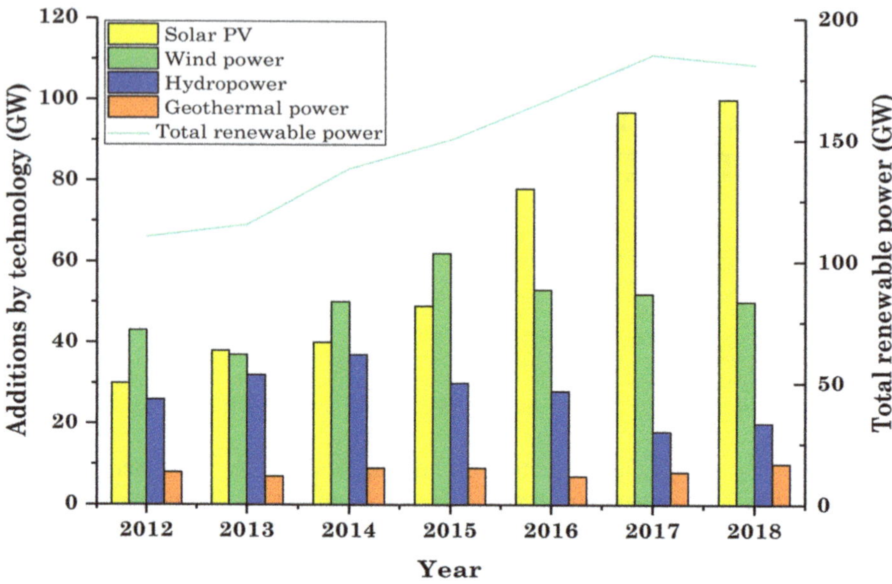

Fig. 3.1 Annual additions of renewable power capacity by various technology

and wind power. Hence, the maximum utilization of RES has become an important issue in all the leading countries in the field of green energy. Therefore, a commitment to immediate actions combined with studies related to optimal operation of RES is necessary today to prevent the waste of renewable energies. In this regard, PowerFactory has great potential to analyze the effects of renewable sources on the power grid. According to conducted research, the exploitation of RES in the power system with consideration of different critical situations has attracted much attention from the researchers' perspective. For example, in [2–4], the main goal is to maximize the utilization of RES in the power grid using energy storage systems to study long term reliability of the system via DIgSILENT PowerFactory. In [5–8], the effect of wind farms was considered to enhance the power grid's stability, compensate power deficit during peak periods, and decrease the power network's operating cost. In these works, all analyses are made by PowerFactory.

The various technologies include photovoltaic (PV) system, wind turbine, and fuel cell can be modeled using two methods in PowerFactory. As shown in Fig. 3.2, it is possible to implement these two methods using the drawing toolbar.

I. **Specific templates**: In this mode, users can utilize wind turbines, PV systems, and fuel cell sources. After selecting each of these sources, the necessary conditions must be provided for connection and operation of RES in the form of the power system. One of the most important things is the use of a suitable transformer to adjust voltages in the power network's busbars.
II. **General templates**: In this mode, users can apply wind turbine in doubly-fed induction generator (DFIG) mode, variable rotor resistance system, or fully rated converter technology. Use this mode also gives users access to the PV

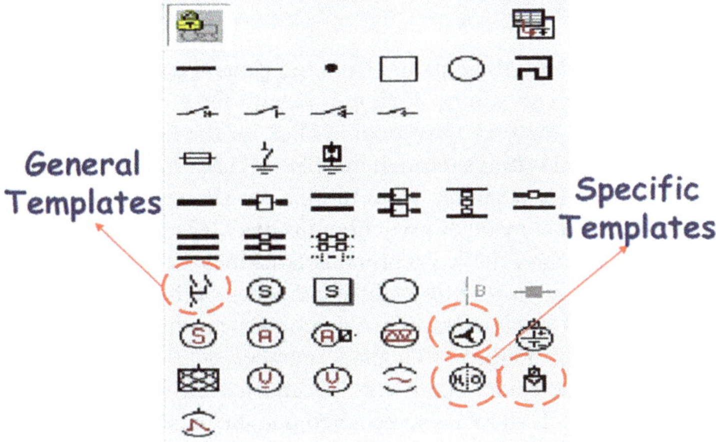

Fig. 3.2 Paths to access renewables in PowerFactory

systems and battery with frequency control. The main advantage of this method compared to the first mode is that in this method all the necessary equipment, including suitable transformer design for connecting RES to the power grid, will be added to the system's graphic by the software. To use this mode, follow the steps below:

- *Select the general templates option from drawing tool*;
- *From the opened window, select the desired RES (Note that the displayed window should not be closed)*;
- *Then drag the mouse into the main window to display the renewable source's symbol*;
- *Left-click on the main window and attach the displayed schematic to the desired busbar. Finally, right-click to end this task.*

Due to the enormous importance of wind and solar PV resources, the required setting and significant tools for these technologies are completely described in the following sub-sections.

3.1.1 Wind Turbine in PowerFactory

In the power system, the role of wind turbines can be examined from different perspectives. However, it is important to note that in order to perform different analyses, we must define a suitable model for the wind turbine. Improper use of RES will not only improve the performance of the power system but will also bring various negative effects to the grid performance. Hence, the settings for the different sections of the wind turbines are described in the following sub-sections.

3.1.1.1 Essential Settings

After selecting the wind turbine model from the general templates provided by the software, basic settings should be done in line with the goals of the power system operator. To this end, the user must double-click on the desired wind turbine and then apply the required settings through the “Basic Data” tab. Important sections of the “Basic Data” tab for a sample wind turbine are shown in Fig. 3.3. Generators used in wind turbines are made of asynchronous machines and normally tend to use reactive power. Therefore, the asynchronous machines’ settings must be properly applied to prevent the voltage drop in different zones of the power grid. According to Fig. 3.3, structural parameters of asynchronous machines, including nominal speed, nominal power, nominal frequency, number of pole pairs, zero sequence impedance, and torque speed curve can be adjusted through the “Type” section (marked ❶ in Fig. 3.3). It should also be noted that the number of parallel machines can be changed from the “Basic Data” tab, depending on the need of the network operator (marked ❷ in Fig. 3.3).

In order to use asynchronous machines in the form of wind turbines, the type of machine must be in generator mode (marked ❸ in Fig. 3.3). Wind turbines can be operated in three modes of the standard asynchronous machine, DFIG, and variable rotor resistance (marked ❹ in Fig. 3.3).

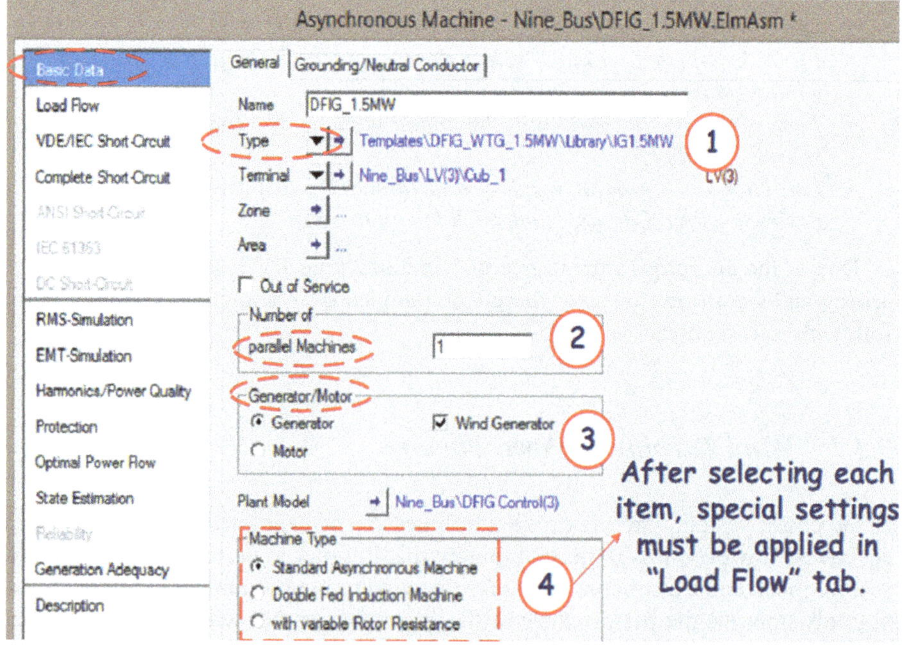

Fig. 3.3 Required settings to launch wind turbines

I. **Standard asynchronous machine**: In this mode, users could determine the busbar's settings and the amount of active power injected to the power grid through the "Load Flow" tab. Users will only be able to control the injected active power into the grid by selecting the "AS" node. However, by selecting the "PQ" node, the user will be able to control both the injected active power as well as the injected/absorbed reactive power. Regarding reactive power support, it should be noted that the positive values represent the injected reactive power into the grid and the negative values indicate the absorbed reactive power from the grid. Therefore, it is advisable to always use the "PQ" mode in the standard asynchronous machine case.

II. **Doubly fed induction machine (DFIG)**: By using this mode, the user will have more options than the standard asynchronous machine mode to apply advanced settings. In addition to the mentioned settings in the first case, it is possible to control the voltage level or the amount of voltage drop (marked ❶ in Fig. 3.4). It is also possible to use a special capability curve for each asynchronous generator (marked ❷ in Fig. 3.4).

III. **Variable rotor resistance**: By applying this case, in addition to the existed settings in the first case, it is possible to control the machine's slip. Induction machine slip can be determined in either fixed or variable mode with active power. The use of variable mode can have a significant impact on transient state studies of large-scale power systems.

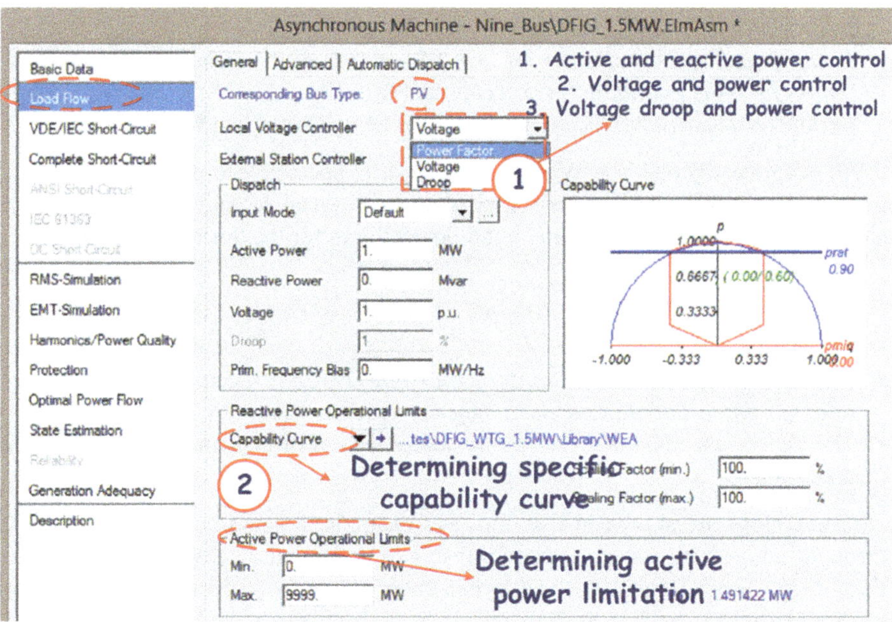

Fig. 3.4 Required settings in DFIG mode

3.1.1.2 Probabilistic Model

In wind turbines, wind speed is known as a random parameter. For this reason, wind power generation is variable throughout the day. The best way to model wind speed variations is to use probability distribution function (PDF). The Weibull PDF is one of the most widely used distribution function for modeling wind speed changes, which is defined by Eq. (3.1) [9, 10]:

$$h(v) = \frac{k}{c}\left(\frac{v}{c}\right)^{k-1} e^{-\left(\frac{v}{c}\right)^{k}} \tag{3.1}$$

where v is the wind speed, k is the shape parameter, and c is the scale parameter of the distribution. According to the international standard IEC 61400-12-1 [11], the best value for the shape parameter (k) is 2. In this case, the Weibull distribution function is converted to the Rayleigh distribution.

To use probabilistic models, it must first be selected through specific templates or general templates available in the drawing toolbar and connected to the desired busbar. Then, to define a new probabilistic model based on the Weibull PDF for wind turbines in PowerFactory, users must follow the steps below:

- *Select the general templates option from drawing tool*;
- *Go to the "Generation Adequacy" tab and active "Wind Model" box* (marked ❶ in Fig. 3.5);
- *In the "Wind Model" box, select the "Stochastic Wind Model (Weibull Distribution)" option* (marked ❷ in Fig. 3.5);
- *In the "Weibull Distribution for Wind Speed" box, select the "Mean & Beta" option from the available section* (marked ❸ in Fig. 3.5). *Then specify the shape parameter (Beta), and the scale parameter (Mean)* (marked ❹ in Fig. 3.5).

To accurately model the Weibull method, it is necessary to determine the wind power curve as well as the meteorological information correctly. The following steps should be performed to assign the wind power curve for the desired wind turbine:

- *Go to the "Generation Adequacy" tab*;
- *Select "New Project Type" from the "Wind Power Curve" section* (marked ❶ in Fig. 3.6);
- *In the opened window, right-click on the "Power Curve" box then select "Append n Rows." After that allocate the number of needed rows* (marked ❷, ❸ in Fig. 3.6);
- *The power-speed curve can be imported into PowerFactory using the created table.*

These steps are shown in Fig. 3.6. For example, a sample power-speed curve through the described steps is shown in Fig. 3.7. As can be seen from this figure, through the "Base Data" tab, the user can enter power-speed curve information into the created table (marked ❶, ❷ in Fig. 3.7). In addition, using the approximation option, users can choose one of the five available methods in this part to draw the curve depending on their needs (marked ❸ in Fig. 3.7). The values for the active power can also be applied in MW or pu, which can be accessed through the

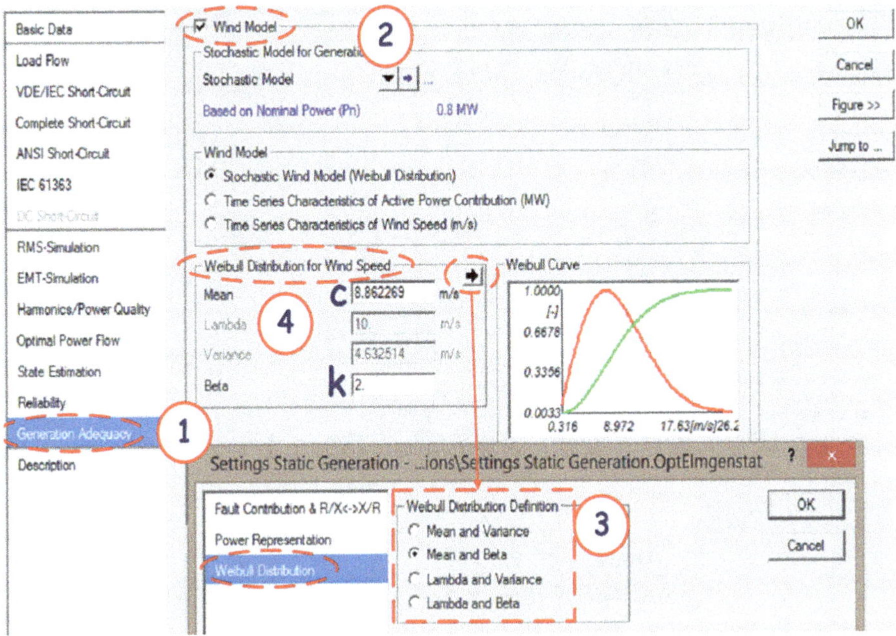

Fig. 3.5 Process of creating probabilistic model (Weibull model)

"Configuration" tab (marked ❹ in Fig. 3.7). This curve is plotted for a default condition. In practice, the wind speed curve related to the particular case study should be applied to the software. It should also be noted that the wind speed curve in an area is always related to different seasons of the year.

In some studies, the effects of using wind farms on the power grid may be assessed. Under such conditions, probabilistic coefficients are usually assumed to be identical for all wind units. Hence, to avoid any difficulty in evaluating the performance of wind farms, a unique factor called "Meteo Station Correlation" is used to distinguish the behavior of the wind farms.

This coefficient varies for different regions and it can be determined from the nearest weather station. In addition, the "Meteo Station" coefficient range in the PowerFactory is from 0 to 229, and to achieve these coefficients, various actions like Fig. 3.8 must be taken.

- *Double-clicking on the desired wind turbine*;
- *Go to the "Generation Adequacy" tab* (marked ❶ in Fig. 3.8);
- *Choose the "Select" option from the "Meteo Station" section* (marked ❷ in Fig. 3.8);
- *In the opened window, select the "New Objective" button in the "Meteo Stations" folder* (marked ❸ in Fig. 3.8);
- *Choose the considered coefficient from the drop-down list* (marked ❹ in Fig. 3.8).

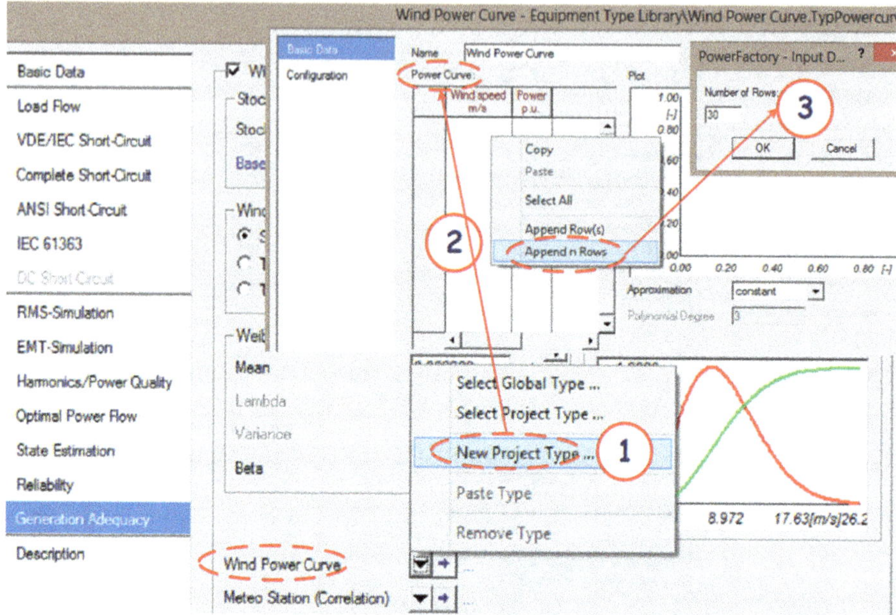

Fig. 3.6 Process of determining power-speed curve for wind turbine

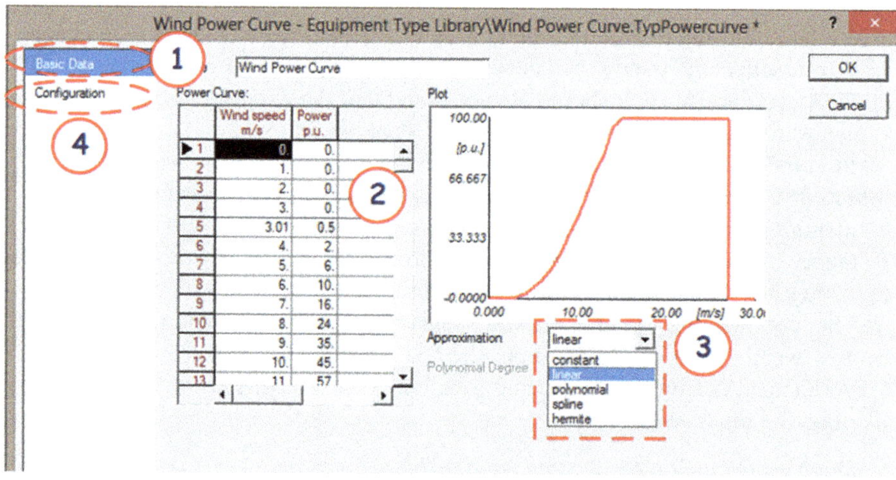

Fig. 3.7 Power-speed curve in PowerFactory

3.1.1.3 Time Series Model

In addition to the probabilistic model, we can model the behavior of wind turbines using time series. In this regard, the decomposition model, autoregressive moving average (ARMA) model, back propagation neural network (BPNN) model, and

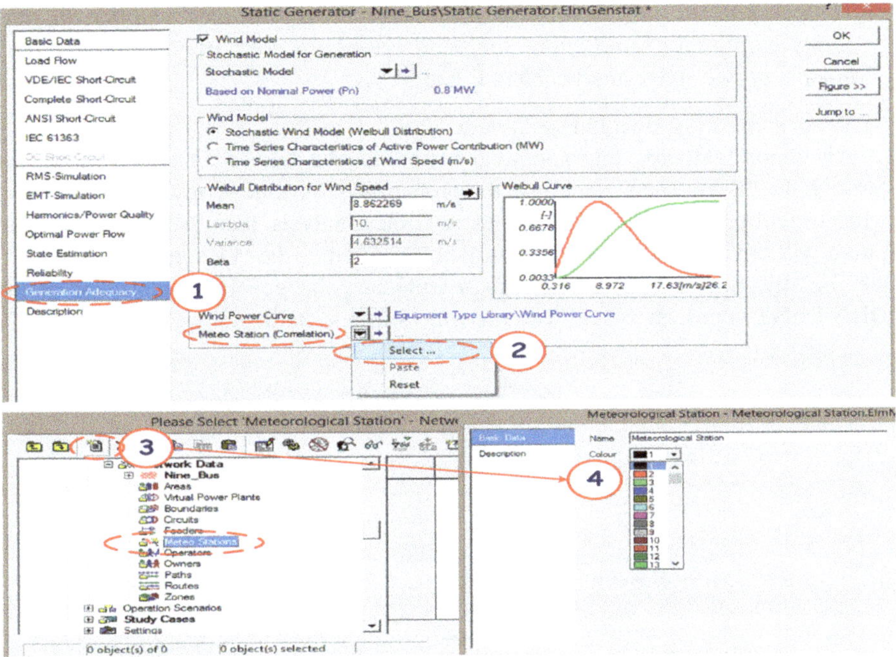

Fig. 3.8 Process of determining Meteo Station coefficients

triple exponential smoothing (TES) series can be used to model the behavior of wind systems [12, 13]. Using prevalent time series, wind speed or produced power of wind turbines can be predicted in different intervals (hourly, monthly, or seasonally). After performing the mathematical calculations, the time series' outputs can be applied to the PowerFactory. Follow these steps to implement this plan:

- *Go to the "Generation Adequacy" tab and active "Wind Model"* (marked ❶ in Fig. 3.9a);
- *In the "Wind Model" box, select the "Time Series Characteristics of Active Power Contribution" or "Time Series Characteristics of Wind Speed" options* (marked ❷ in Fig. 3.9a);
- *Users can change the number of studied years through "Number of Sample" box* (marked ❸ in Fig. 3.9a);
- *Double-clicking on the time series box* (marked ❸ in Fig. 3.9a);
- *In the opened window, go to the "Operational Library" and select the "New Object" button in the "Characteristics" folder. Then define new parameter characteristic* (marked ❹, ❺ in Fig. 3.9a);
- *Go to the "Scale" option and select "New Object" button in the "Characteristics" folder. Then define time scale element* (marked ❶–❸ in Fig. 3.9b);
- *In the pop-up window, select the desired time interval (day, month, or year) and enter the related values using "Append n Rows" option, e.g., for "month of year" unit, the user must allocate twelve rows per year* (marked ❶–❸ in Fig. 3.9c);

- *Then go back to the "Time Series Characteristics" window and enter the time series outputs for wind speed (or wind power) in each row (an indicator of months of one year)* (marked ❹ in Fig. 3.9c).

These steps are shown in Fig. 3.9a–c.

It is important to note that a nominal power must be defined for each wind turbine through the "Basic Data" tab to use the probabilistic or time series functions. Then, when executing *Load Flow Analysis* or any other analysis, the wind turbine's power rating will change based on the information entered in the "Generation Adequacy" tab. Finally, to better understand these subjects, the general mechanism of the DIgSILENT approach to deal with the wind speed variations is shown in Fig. 3.10.

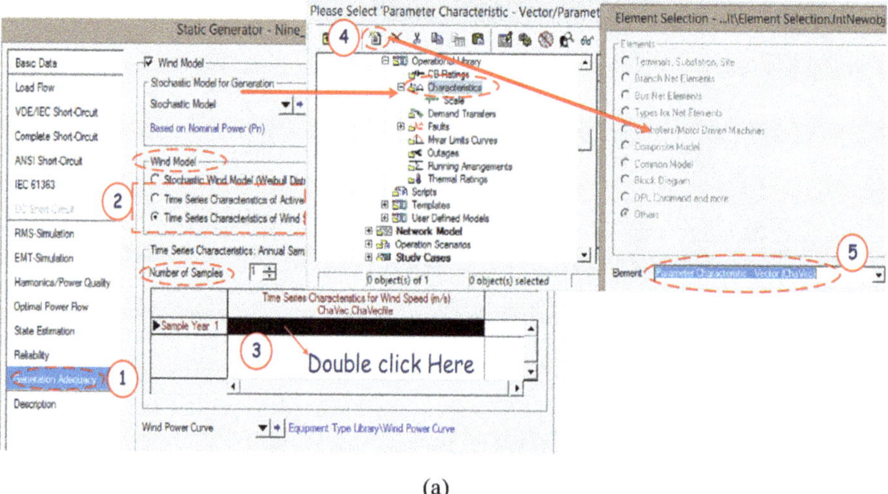

(a)

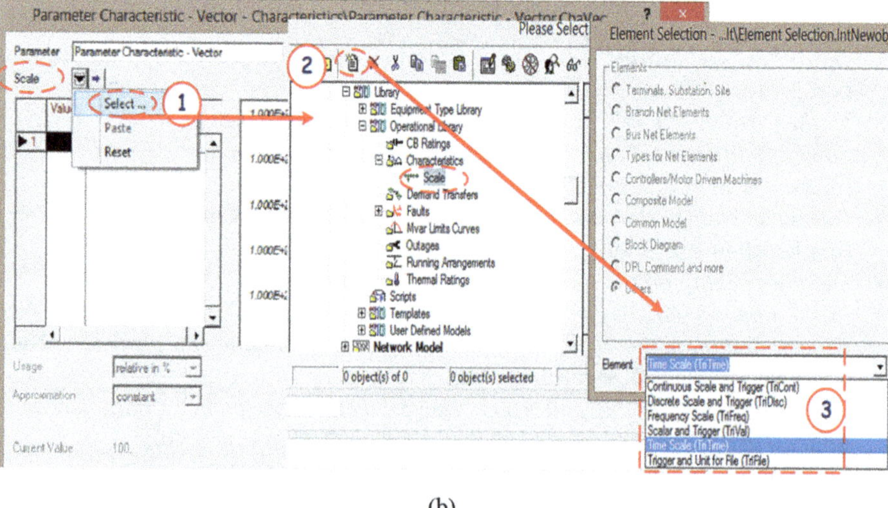

(b)

Fig. 3.9 Process of determining time series characteristics (**a**) defining time series, (**b**) determining the scale of project, and (**c**) adjusting required values

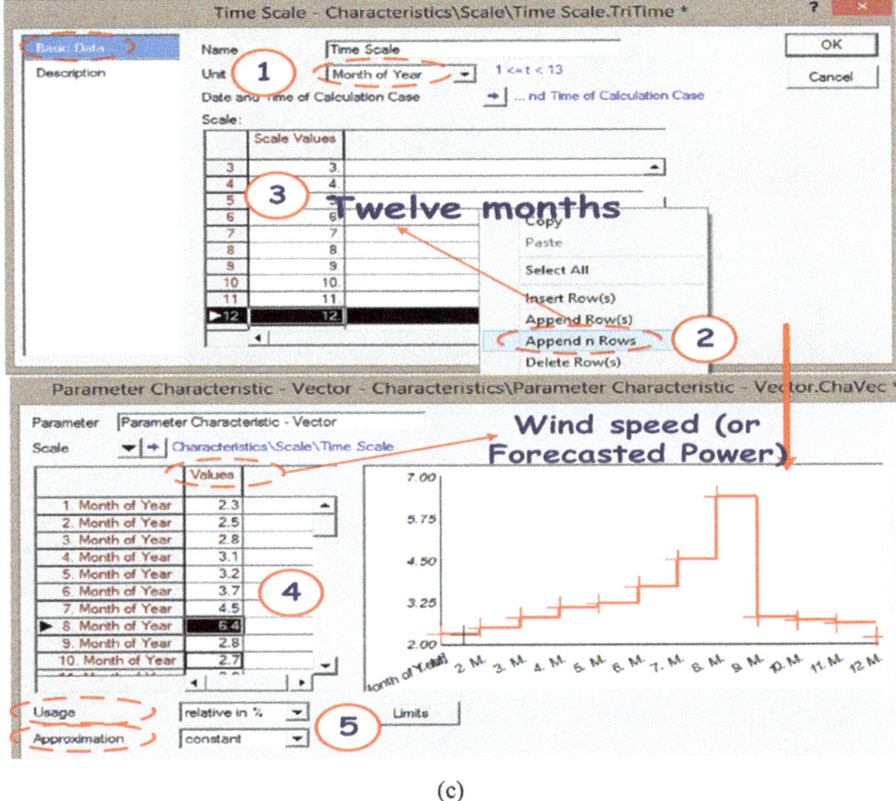

(c)

Fig. .9 (continued)

3.1.2 *Photovoltaic System in PowerFactory*

Like other RES, the performance of photovoltaic (PV) systems is highly dependent on environmental conditions. Solar radiation and ambient temperature are the two most influential parameters on the produced power of PV systems. The efficiency of PV systems also varies depending on the different technologies used by solar cell manufacturers. In this regard, solar cell datasheets are the best way to access all the information about them. The information provided in each datasheet is based on the standard test conditions (solar radiation equal to 1000 W/m^2 and ambient temperature equal to 25 °C). For example, Table 3.1 demonstrates a sample solar cell datasheet, which is manufactured by Yingli Solar Company. The provided data in this table are useful for PV cell simulation at the standard test conditions. However, in reality, it rarely happens that solar systems operate at the standard test conditions. Because solar radiation and ambient temperature can vary from region to region and from season to season. In Table 3.1, the symbols P_{max}, V_m, I_m, V_{oc}, and I_{sc} are the maximum power, optimum operating voltage, optimum operating current, open-circuit voltage, and short-circuit current, respectively. These parameters are strongly

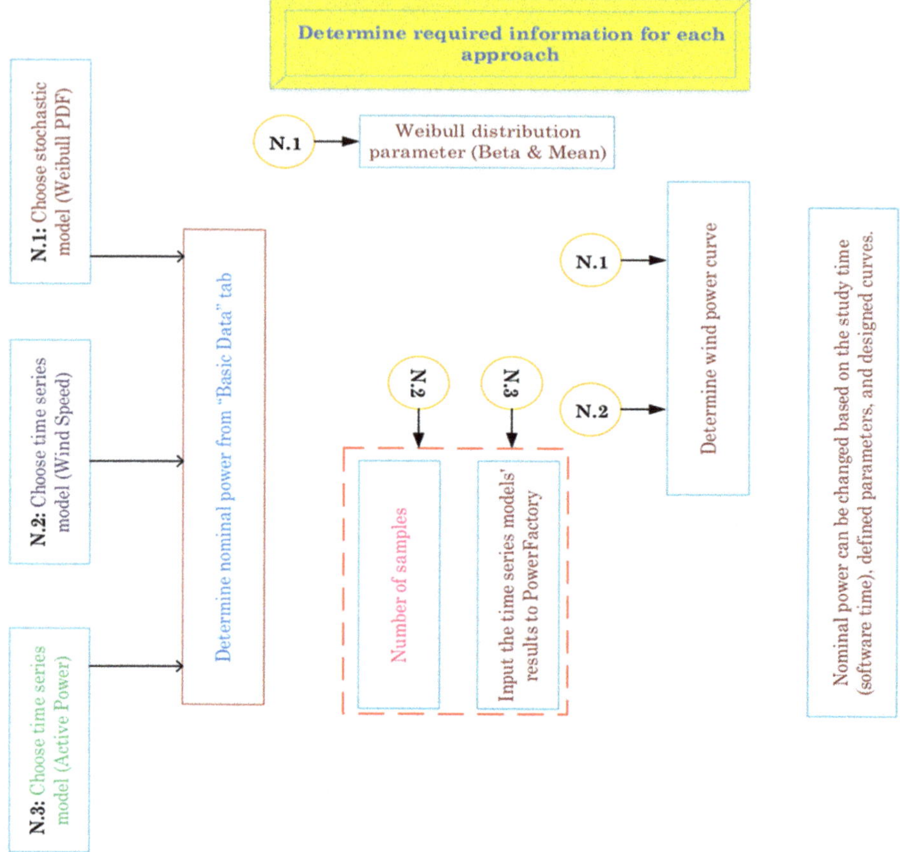

Fig. 3.10 Conceptual architecture of wind turbine modeling in PowerFactory

Table 3.1 Sample solar cell data

Manufacturer	Model	Dimension	Cell material
Yingli solar	YL250C-30b	156 × 156 mm	Mono-crystalline

Electrical characteristics					
Efficiency	P_{max}	V_m	I_m	V_{oc}	I_{sc}
17.5%	250 W	30.96 V	8.07 A	37.92 V	8.62 A

Temperature coefficients of (%/°C)		
P_{max}	V_{oc}	I_{sc}
−0.4	−0.35	+0.06

Correction factors for irradiance		
Solar radiation (W/m^2)	Voltage correction (V/V)	Current correction (A/A)
1000	1	1
800	0.98	0.79
600	0.97	0.59
200	0.91	0.21

influenced by solar radiation and ambient. I–V and P–V curves can be used to show the effect of solar radiation and cell temperature on the output power of the PV system. These curves for YL250C-30b solar cell are shown in Figs. 3.11 and 3.12. These figures show that the highest PV production power is at high solar irradiance and low ambient temperature. In general, Eq. (3.2) can be used to estimate the PV production power under different conditions [14].

$$P_{PV}(t) = A \times \eta_{PV} \times R(t) \times \left[1 - 0.005 \times \left(T(t) - 25\right)\right] \tag{3.2}$$

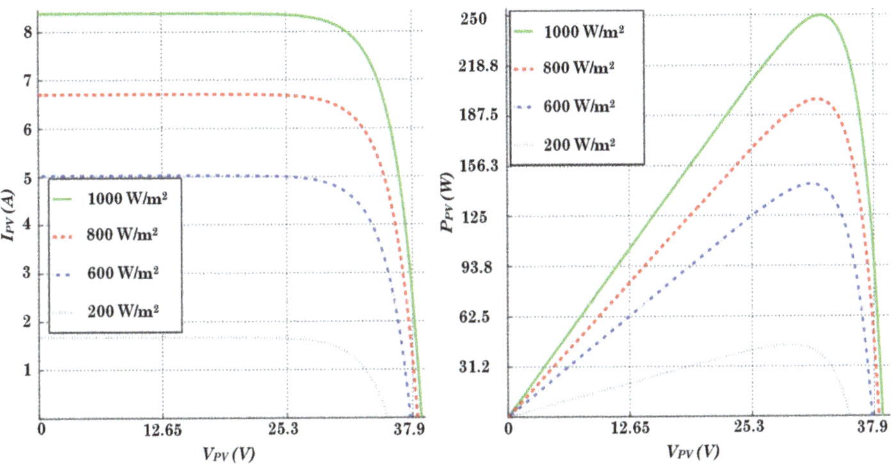

Fig. 3.11 I–V and P–V curves of the PV cell YL250C-30b with constant temperature (25 °C) and variable irradiance

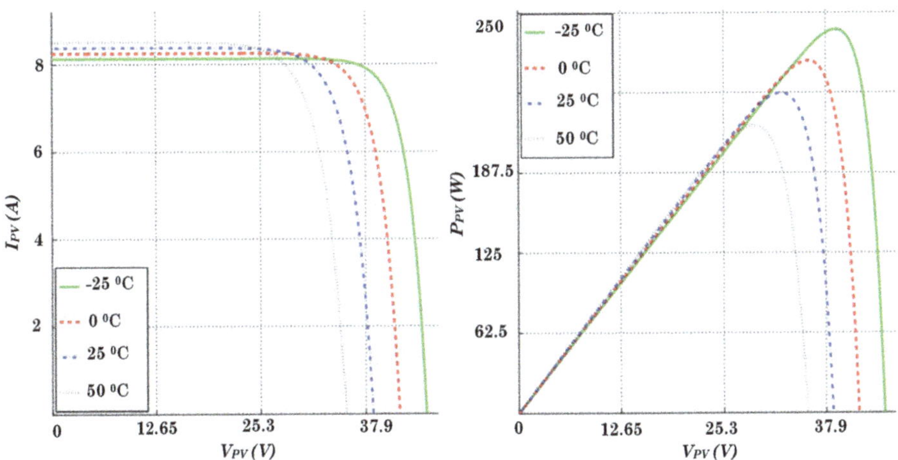

Fig. 3.12 I–V and P-V curves of the PV cell YL250C-30b with constant irradiance (1000 W/m^2) and variable temperature

where, A is the panel area (m^2), $\eta_{PV}$is efficiency of PV system, $R(t)$ is the solar radiation at hour t (W/m^2), and $T(t)$ is the ambient temperature (°C).

3.1.2.1 Active Power Input Mode

In this mode, users can model PV systems based on daily changes in output power. To this end, users must follow the steps below:

- *Double-clicking on the PV system;*
- *Select "Active Power Input" mode from the "Model" section in the "Basic Data" tab;*
- *Go to the "Operating Point" section in the "Load Flow" tab;*
- *Right-click on the "Active Power" box then select "Add Project Characteristic-Time Characteristics";*
- *In the opened window, go to the "Operational Library" and select "New Object" button in the "Characteristics" folder;*
- *After the above steps, a new object from the "ChaTime" class opens automatically;*
- *By using the "Table Data" option, the PV system output power profile can be imported directly into the existing table, as well as by using the "File Data" option. This profile can be imported into PowerFactory through an external Excel file (∗.csv).*

The performance of the different sections in the time characteristics window is explained in detail in Chap. 2. For example, the active power profile for a 10 kW PV system that imported into PowerFactory through the described steps is shown in Fig. 3.13. This curve changes with the different seasons of the year.

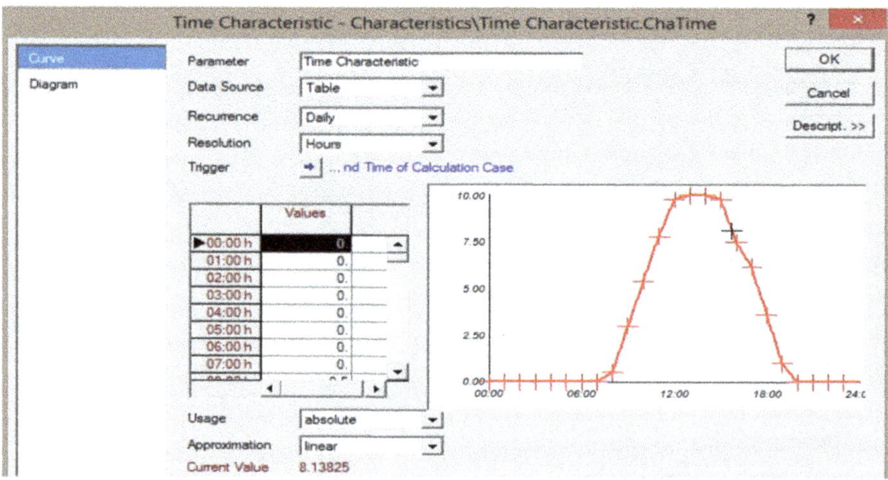

Fig. 3.13 PV output power profile in PowerFactory

3.1.2.2 Solar Calculation Mode

Using this mode, the user can evaluate the effects of structural characteristics of solar cells as well as the environmental conditions on the performance of the photovoltaic systems. The structural characteristics of each solar cell can be obtained through the specific datasheet of that cell. The environmental conditions of each region can also be obtained through geographical maps for different seasons of the year. Information on several reputable manufacturers, such as BP Solar Company, Sharp Solar Company, and Yingli Solar Company is available in the PowerFactory library. According to Fig. 3.14, this information is available for PV systems through the "Select Project Type" option in the "Type" section, "General" header (marked ❶ in Fig. 3.14). Users can also customize peak power, open-circuit voltage, and short circuit current parameters after selecting the desired type. Besides, the three phase or single phase technology for PV systems can be selected from the "Basic Data" tab (marked ❷ in Fig. 3.14). The number of inverters used and the number of panels connected to each inverter can also be determined (marked ❸ in Fig. 3.14). In the "Basic Data" tab, through the second header "System Configuration," users can determine the geographical location of the PV system, orientation angle, tilt angle, and efficiency factor of the inverter.

As already explained, the maximum power set for the PV system is only for standard conditions, 1000 W/m^2, and 25 °C. For the sample PV system (YL250C-30b), this value was 250 W, which was determined by the "Basic Data-Type" section. Users can do their studies on the power grid for different radiation levels as well as different ambient temperatures for 24 h. To this end, much-needed weather information can be set for different hours of the day through the "Environment Data" header in the "Load Flow" tab. The best way to determine the irradiance curve is to use the global and direct radiation options. Irradiance and

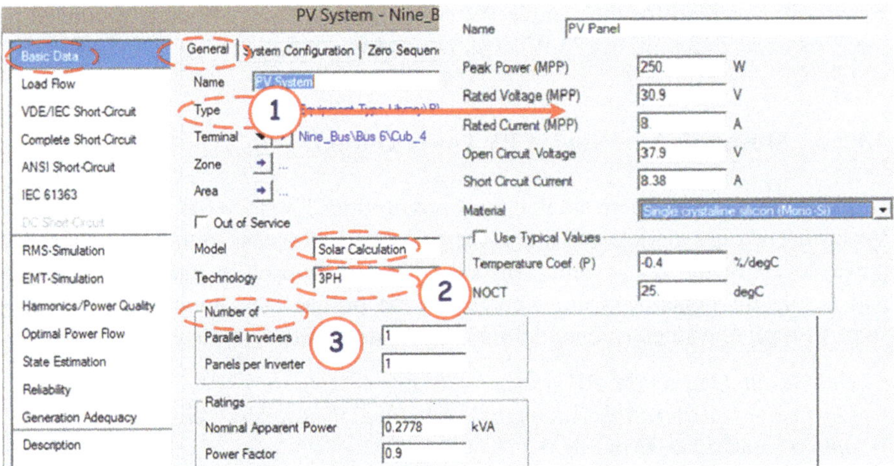

Fig. 3.14 Process of determining the structural characteristics of PV systems

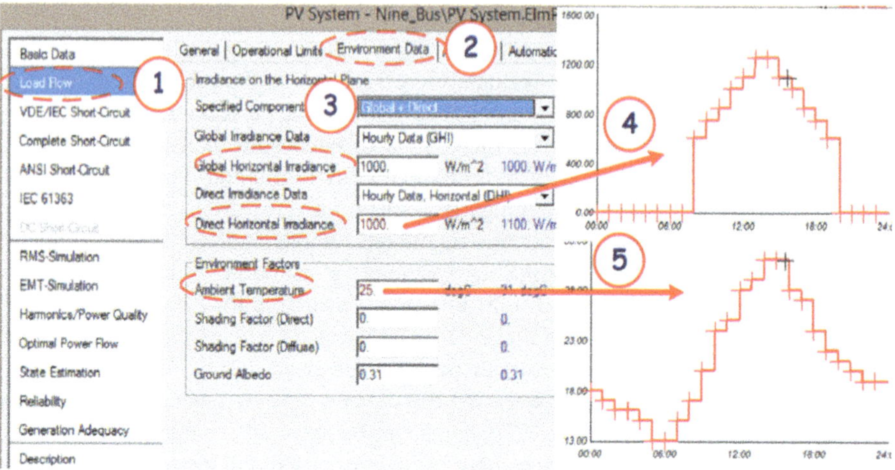

Fig. 3.15 Process of determining the irradiance situation and temperature values

temperature curves can be determined precisely similar to the described steps in the previous sub-section (determining output power profile). These steps are shown in Fig. 3.15.

It is important to note that in the "Solar Calculation" mode, users do not have access to the output power of the PV system. In this mode, the output power of the PV system is determined by PowerFactory according to the irradiance data, temperature values, as well as the structural parameters of each solar cell. In this regard, Eq. (3.2) can help users to estimate the amount of solar power generation under different conditions. In addition, the real-time power generation dispatch by the PV system can be observed from the "General" header in the "Load Flow" tab with respect to the set parameters. According to the temperature and irradiance curves previously specified for the sample PV system, the amount of power produced by the PV system at 3 p.m. is 273 W, as shown in Fig. 3.16.

3.1.2.3 Stochastic Modeling of PV Power Output

Unlike wind turbines, there are not many possibilities for stochastic modeling of PV systems. The only model available is the possibility of access to the different active power levels of the PV system. This model is mostly used in *Reliability Analysis* and the results depend on the accuracy of the probabilistic model. The required steps to apply the stochastic model for a PV system in the power grid include:

- *Double-clicking on the PV system*;
- *Go to the "Generation Adequacy" tab, and then select "Stochastic Model" option* (marked ❶, ❷ in Fig. 3.17);

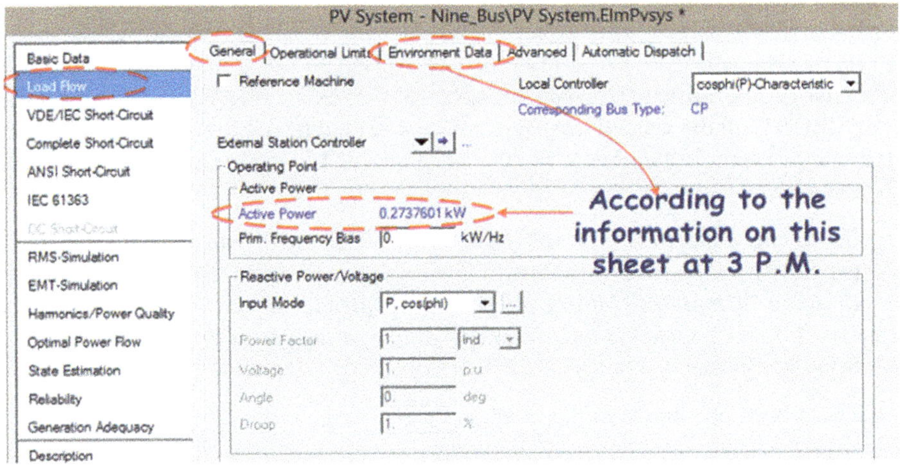

Fig. 3.16 Real-time power generation dispatch by the PV system

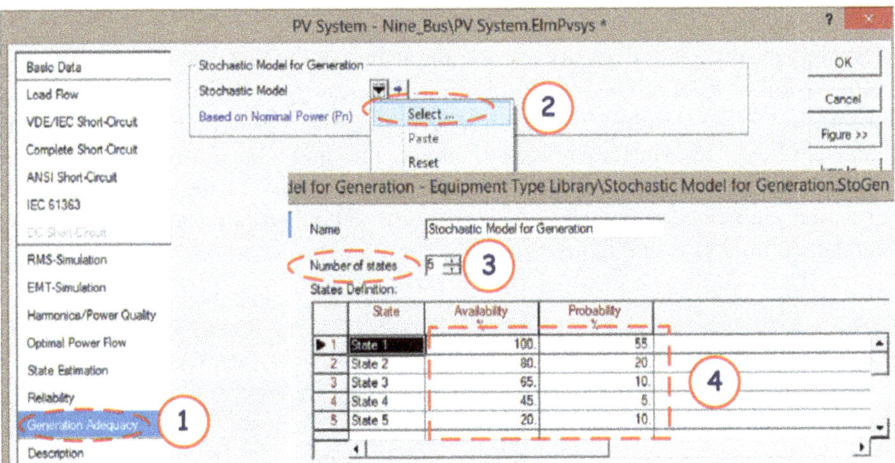

Fig. 3.17 Process of creating stochastic model for PV system

- *In the opened window, the number of probabilistic modes must be specified* (marked ❸ in Fig. 3.17);
- *The probability of availability to each production level must be determined through the "States Definition" box. Note that the sum of probabilities must be equal to 1* (marked ❹ in Fig. 3.17).

These steps are shown in Fig. 3.17.

3.1.3 Adding Elements of Renewable Sources to the Library

To change the structural characteristics of the existed RES templates in the DIgSILENT, all the needed templates must be copied from the software's library to the study case project's library. For this purpose, the following steps should be done by the user:

- *Go to the DIgSILENT library from "Data Manager" window* (marked ❶ in Fig. 3.18);
- *Copy the "Templates" library and paste the copied templates into the "Equipment Type Library," which is located on the active project's library window* (marked ❷ in Fig. 3.18).

These steps are shown in Fig. 3.18.

3.2 Basic Assumptions to Analyze the Behavior of RES in Power Grids

The main purpose of this section is to examine the role of introduced models for the exploitation of RES on the power grid from different perspectives. The power grid under study is the modified 33-bus 12.66 kV radial distribution standard test system, shown in Fig. 3.19. The data related to the line info and other parameters about this test system are available in [15]. To investigate the impact of the activity schedule of various subscribers, the network under study is divided into three sections: residential, commercial, and industrial.

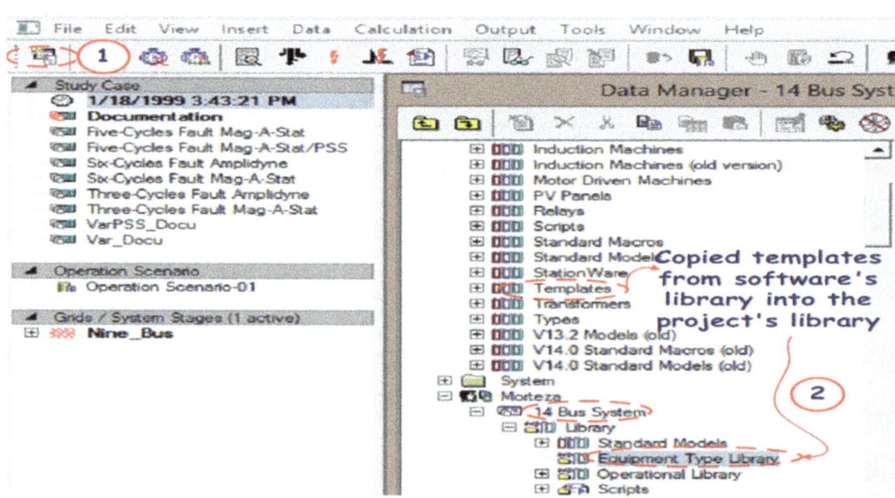

Fig. 3.18 Transfer RES templates from PowerFactory's library to active project's library

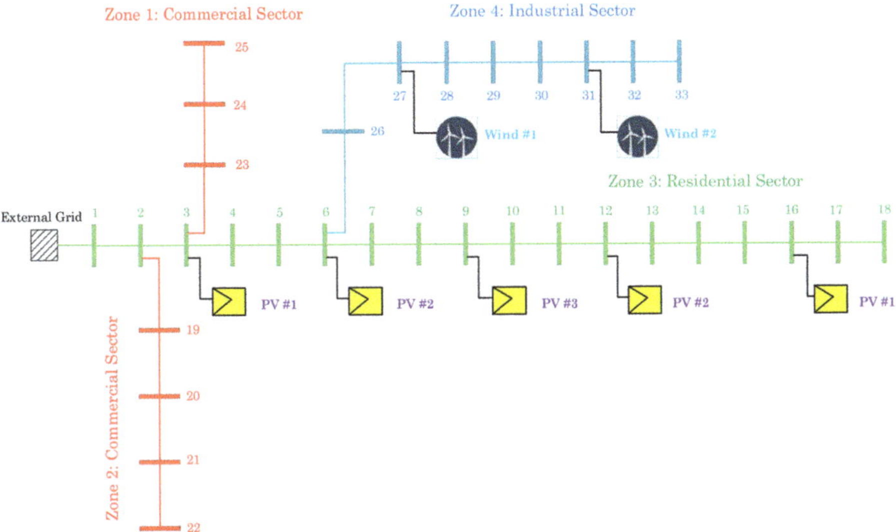

Fig. 3.19 Single diagram of 33-bus network along with RES

The mathematical optimization methods were used in academic discussions to determine the optimal size and location for installing RES. In this regard, much research has been done to minimize power losses or the operating cost of the power grid using meta-heuristic approaches and the General Algebraic Modeling System (GAMS) software [16–19]. In this study, the optimal size and location for the installation of PV systems and wind turbines on the 33-bus test system were determined according to the presented optimization results in [20]. PV systems were used in three types as well as wind turbines in two different types and located at buses 3, 6, 9, 12, 16, 27, and 31. The maximum power capacity of PV systems and wind turbines are equal to 1300 kW (*PV#1*), 2600 kW (*PV#2*), 3200 kW (*PV#3*), 3500 kW (*Wind#1*), 1000 kW (*Wind#2*), respectively. The output power of PV systems and wind turbines that was based on the information provided in previous sections are shown in Figs. 3.20 and 3.21. In addition, in this study, three types of time-varying load models are considered for residential, commercial, and industrial subscribers, which are shown in Fig. 3.22.

Distribution networks can be analyzed from different points of view, such as the different bus voltage levels, active power losses in different feeders, as well as lines loading level. For this purpose, at first, the status of the test system is examined without the presence of RES. In this regard, *Quasi-Dynamic Simulation* is the best analysis tool for evaluating the performance of the power grid over a specific period. The features and required settings to use this tool to analyze the power grid are described below.

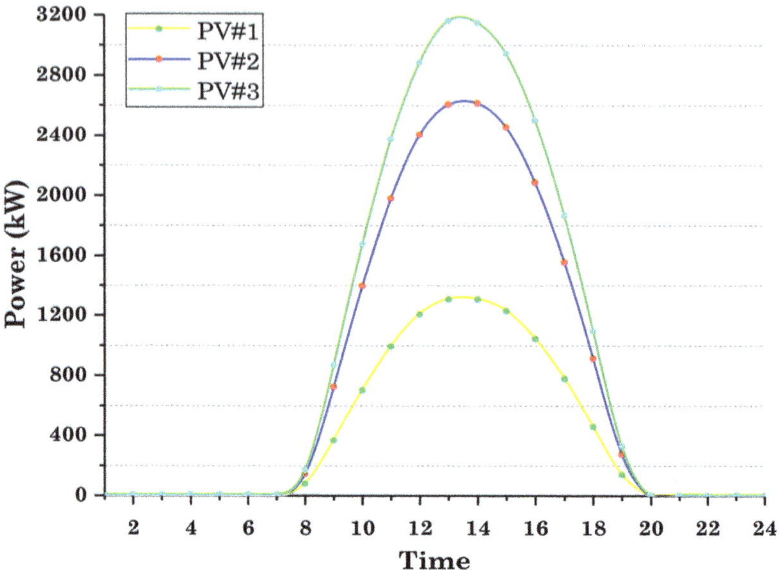

Fig. 3.20 PV systems power output

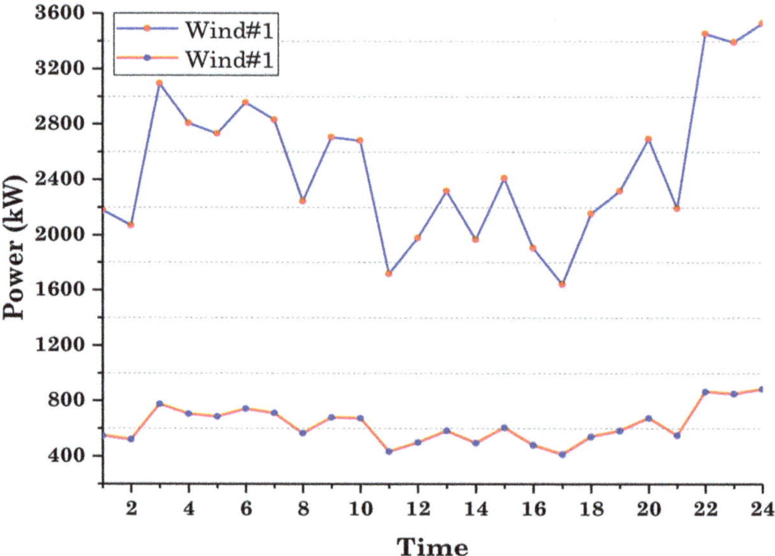

Fig. 3.21 Wind turbines power output

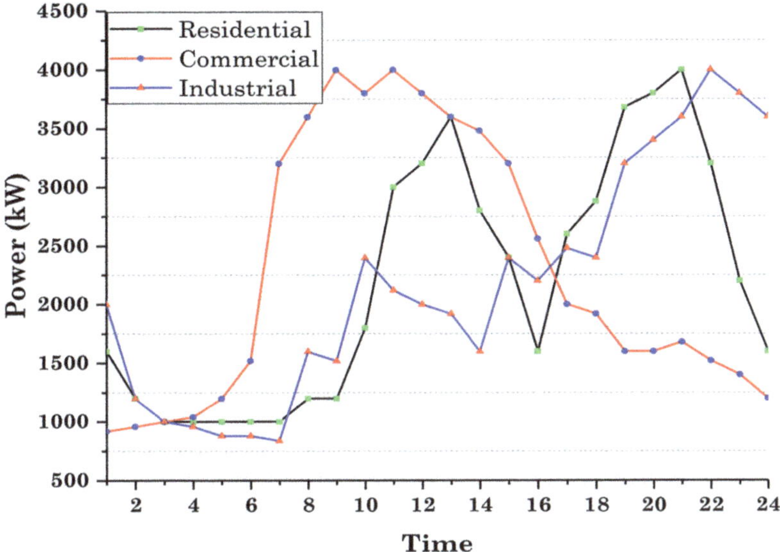

Fig. 3.22 Daily demand profile for each subscriber

3.3 Quasi-Dynamic Simulation

The network operator always tries to identify the various factors affecting the power grid performance and evaluate the network characteristics under different conditions. The various loading conditions can be affected by changes in active power load, power output from RES, maintenance schedule, and unscheduled outages. In this regard, the power grid's operator can analyze the behavior of the system at a specific interval with different time-steps by using the *Quasi-Dynamic Simulation*. Generally, this simulation consists of four steps:

Step 1) Determining the time characteristic of loads and generation resources:

Determining the time characteristics for the system's loads as well as the RES is one of the prerequisites to run this analysis in the PowerFactory. In the previous sections, how to set time characteristics for the system's loads and renewable sources are fully described. However, it is virtually impossible to determine the time characteristics for the various elements in large-scale power systems. In large-scale power systems, users can go through the following steps to determine time characteristics:

- *Initially, according to* Fig. 3.23, *one or more of the time characteristics must be defined for the active power related to the network loads and the produced power of RES.*
- *Press the "Edit Relevant Objects for Calculation" button located on the main menu and select "General Load" icon or RES*;

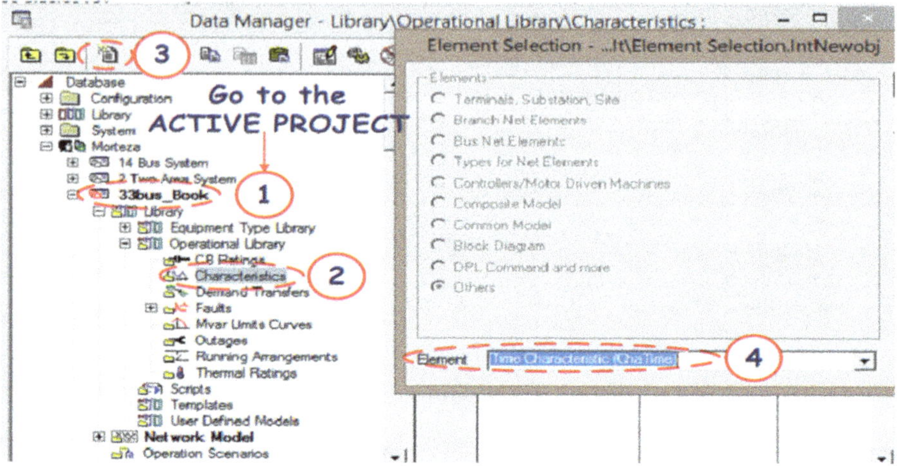

Fig. 3.23 Process of determining a global time characteristics

- *In the pop-up window, go to the "Load Flow" tab*;
- *Highlight the "Active Power" column for all the desired elements to assign time characteristic*;
- *Right-clicking on the highlighted column and then choose "Time characteristics" option from the "Characteristics-Add project characteristics" menu*;
- *Assign the desired characteristic for all the selected elements.*

Step 2) Defining the required output variables for network monitoring and analysis:

By default, various parameters are defined for some elements of the power system to be displayed in the simulation output. As shown in Fig. 3.24, the following steps must be followed to access these elements:

- *Click the "Edit Results Variable" button* (marked ❷ in Fig. 3.24);
- *Select the type of desired results based on the load-flow analysis, which includes AC, AC unbalanced or DC* (marked ❸ in Fig. 3.24);
- *By default, it is possible to monitor outputs for some elements such as grid, transformer, lines, and feeders. To modify the default variables, go to the variable browser dialogue by double-clicking on each element available in this dialogue* (marked ❹ in Fig. 3.24);
- *In the pop-up window, go to the "AC/DC Quasi-Dynamic Simulation" and then select variables from the desired variable set*;
- *Users can use the "New Object" button to define new elements* (marked ❺ in Fig. 3.24);
- *In the pop-up window, enter the object class in the "Class Name" box. All elements have a specific class name. For example, the PV system class name is "ElmPvsys"*;
- *Press "tab" key to update the dialogue and then select the desired variable set for the defined element through the desired page.*

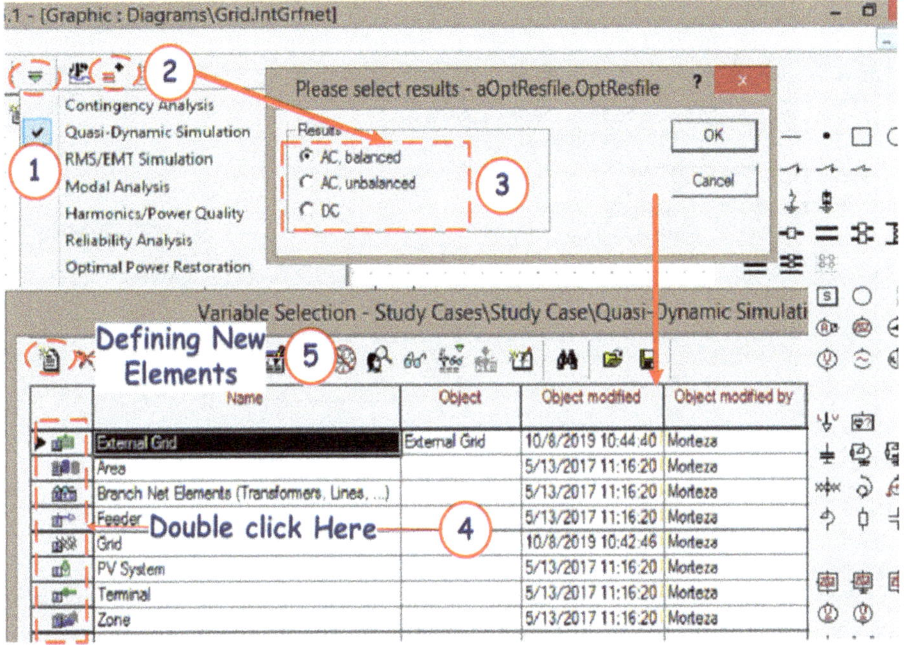

Fig. 3.24 Process of defining the variables for monitoring Quasi-Dynamic simulation

Another way to monitor each element of the power grid is as follows:

- *Right-clicking on the desired elements*;
- *Select "Results for Quasi-Dynamic Simulation" option from the "Define" section.*

Step 3) Running the simulation and obtain required outputs:

The *Quasi-Dynamic Simulation* performs a number of *Load Flow Analysis* for a user-defined interval and displays the results based on the defined variables. To perform the *Quasi-Dynamic Simulation*, press the "*Calculation*" button located on the main menu, and select *Quasi-Dynamic Simulation*. Then the settings dialogue will pop-up, which contains the settings for time intervals. The required settings to run Quasi-Dynamic simulation correctly are as follows:

I. **Load Flow tab**: Users can fulfill the requirements for performing *Load Flow Analysis* from this section (marked ❷ and ❸ in Fig. 3.25).
II. **Time Period tab**: The period of time that users intend to analyze the performance of the power system is determined in this section. The timeframe can range from several hours to several years (marked ❹ in Fig. 3.25).
III. **Sep Size tab**: The size of the required steps to run the simulation in the determined interval is specified by this section. Time steps can be adjusted from one second to several years according to the type of network and required information. Just note that the entered value in the "Step" box must be a positive integer (marked ❺ in Fig. 3.25).

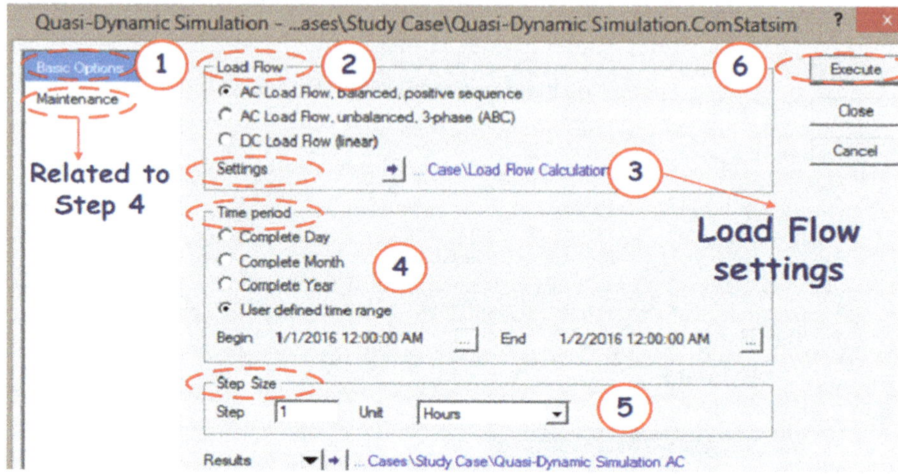

Fig. 3.25 Process of performing Quasi-Dynamic simulation

According to the defined variables, the results of the *Quasi-Dynamic Simulation* can be observed to analyze the power grid performance. There are two ways to view simulation outputs: (1) time diagrams of output variables via the "Create Subplot" icon in the *Quasi-Dynamic Simulation* menu, (2) reports of outputs variables in the form of different tables via the "*Quasi-Dynamic Simulation Reports*" icon in the simulation menu.

The required steps to evaluate the system's elements in the form of a graphical illustration are described in the following:

- *Click the "Create Subplot" icon* (marked ❶ in Fig. 3.26);
- *Go to the "y-Axis" tab* (marked ❷ in Fig. 3.26);
- *Determine the desired element through double-clicking on the "Element" box* (marked ❸ in Fig. 3.26);
- *Determine the desired variables through double-clicking on the "Variable" box* (marked ❹ in Fig. 3.26);
- *Go to the "x-Axis" tab and determine "x-Axis Variable" range*;
- *Go to the "Advanced" tab and specify different indexes, such as chart type (Curves or Bars) or axis description.*

These steps are shown in Fig. 3.26. In addition to graphical diagrams, it is possible to see simulation outputs in the form of different tables for loading range and voltage range of the defined elements. To do this, at first, select the "*Quasi-Dynamic Simulation Reports*" icon and then should be applied the required settings to view the outputs. The available tools to access text reports are shown in Fig. 3.27.

Step 4) Defining the planned outage for maintenance or various faults:

Because of annual repairs, some of the equipment in power systems must be disconnected for some time. Therefore, the effects of different equipment outages should be investigated on network operation as well as other elements. For this purpose, *Quasi-Dynamic Simulation* capabilities can be used to investigate the role of

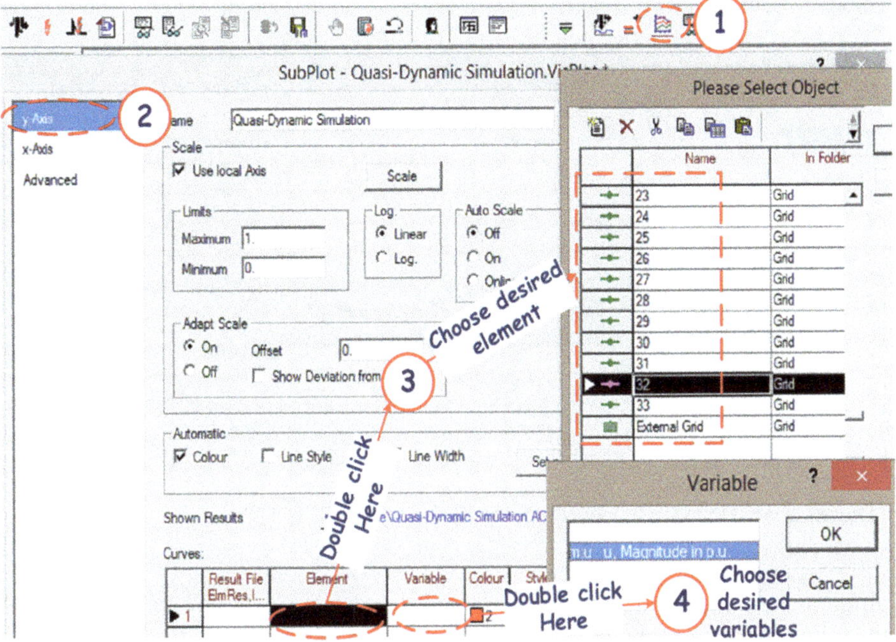

Fig. 3.26 Tips on drawing Quasi-Dynamic analysis outputs

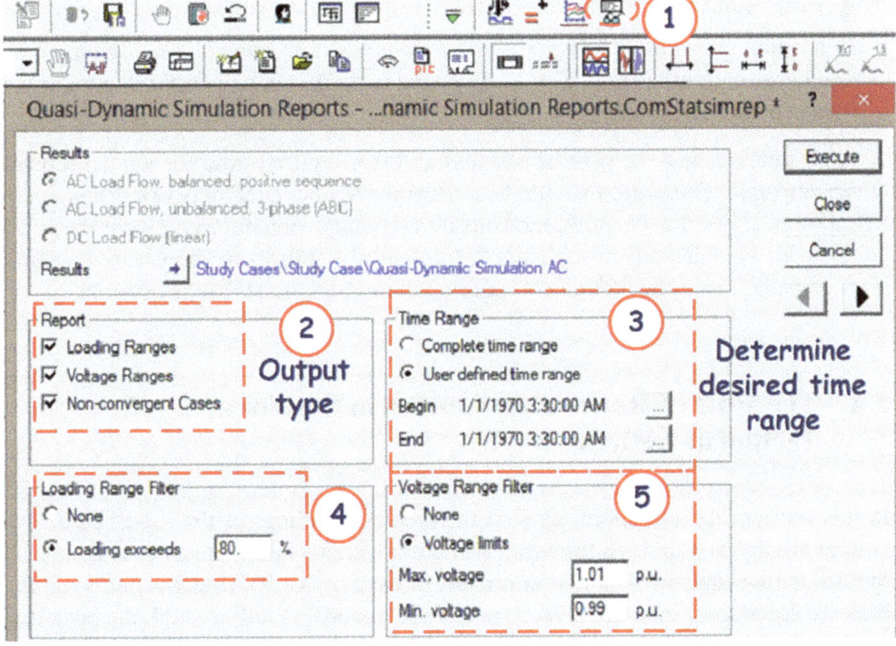

Fig. 3.27 Tips on accessing Quasi-Dynamic analysis text reports

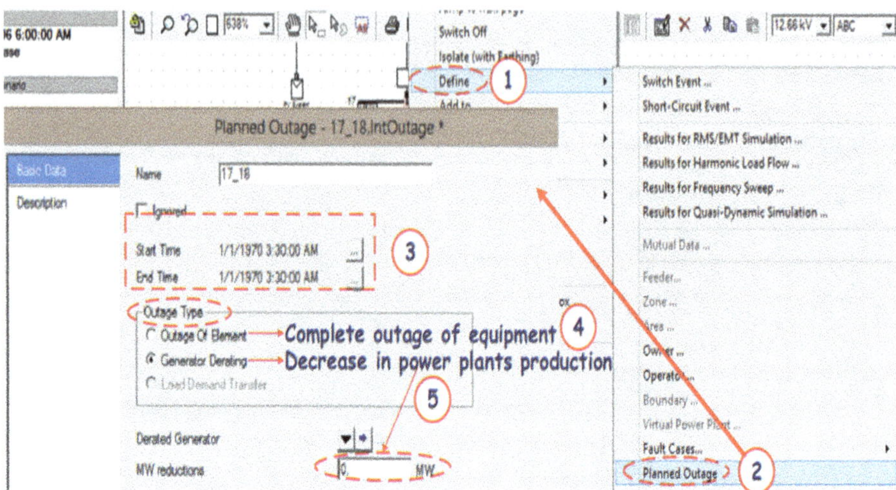

Fig. 3.28 Process of determining the outage times of various elements

different equipment outages in power system performance. According to Fig. 3.28, the following steps must be taken to achieve this goal:

- *Right-click on the desired element then choose "Planned Outage" from "Define" option* (marked ❶, ❷ in Fig. 3.28);
- *Determine the outage time of the element as well as its reactivation time* (marked ❸ in Fig. 3.28);
- *Determine the outage type for desired elements. The complete outage for lines and busbars* (marked ❹ in Fig. 3.28), *or the reduction in power plants production* (marked ❺ in Fig. 3.28);

After determining the type of element and the required time for annual repairs, *Quasi-Dynamic Simulation* should be performed to check the network status in different modes. For this purpose, the considered outage element to perform the simulation must be selected through the "*Maintenance*" tab from the Quasi-Dynamic settings page. From the "*Show all*" option, all defined modes are accessible.

3.4 The Role of Renewable Sources in System's Technical Indices

In this section, the effects of RES on the technical indices of the modified 33-bus system are investigated. To this end, the network status in the presence and absence of RES is investigated. The maximum power capacity of PV systems and wind turbine are equal to 11 and 4.5 MW, respectively. According to Fig. 3.29, the peak load of the test system occurred at 9 p.m. and is equal to 95.49 MW. Therefore, the

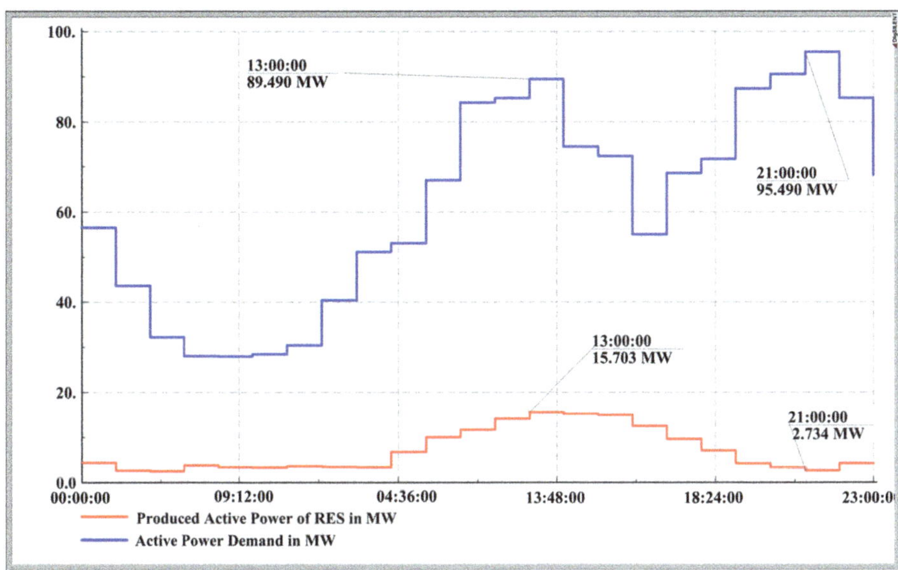

Fig. 3.29 Active power demand and produced power of RES

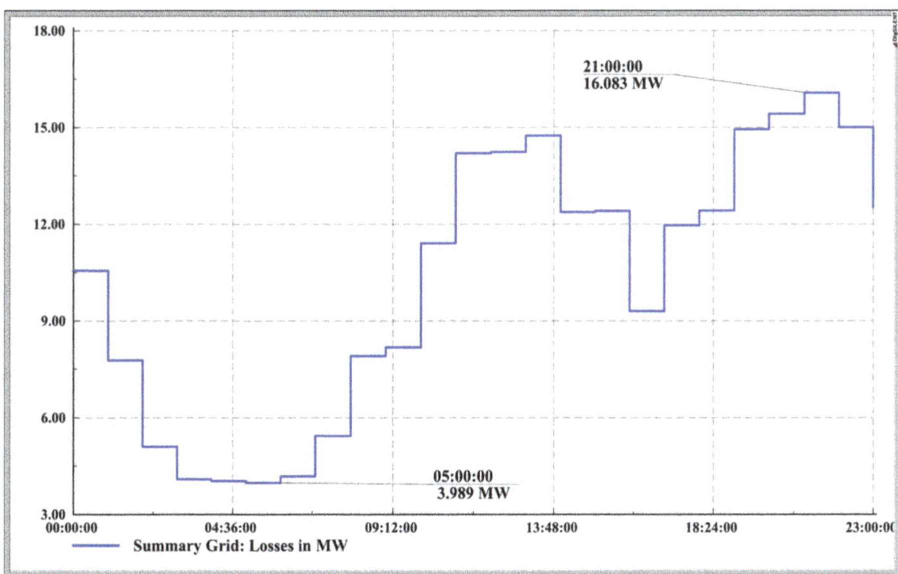

Fig. 3.30 Active power losses during 24 h regardless of RES

maximum system loss is 16.083 MW, which occurred at 9 p.m. This concept is illustrated in Fig. 3.30. The total active power losses of the 33-bus test system during 24 h are 246 MW. As shown in Fig. 3.31, the total amount of system losses has been reduced to 175 MW after applying all RES.

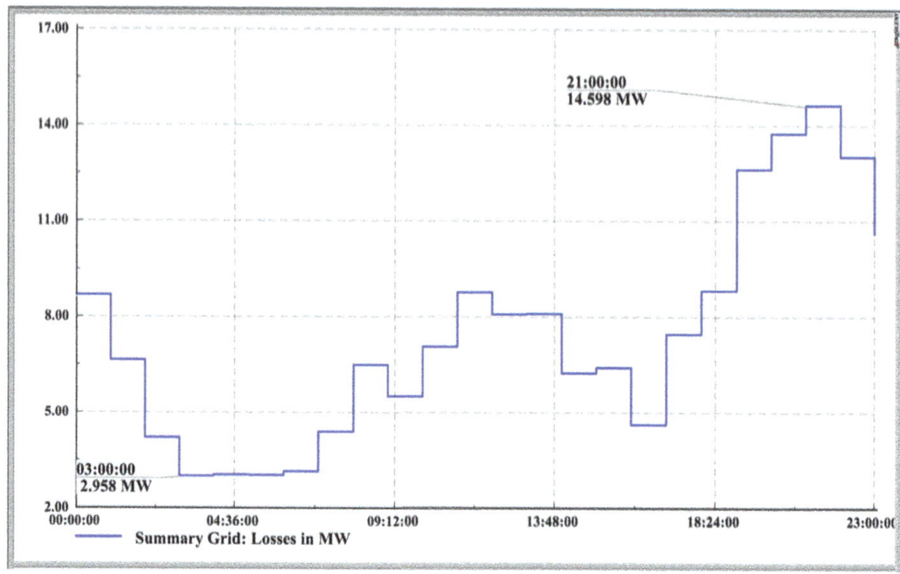

Fig. 3.31 Active power losses during 24 h with considering RES

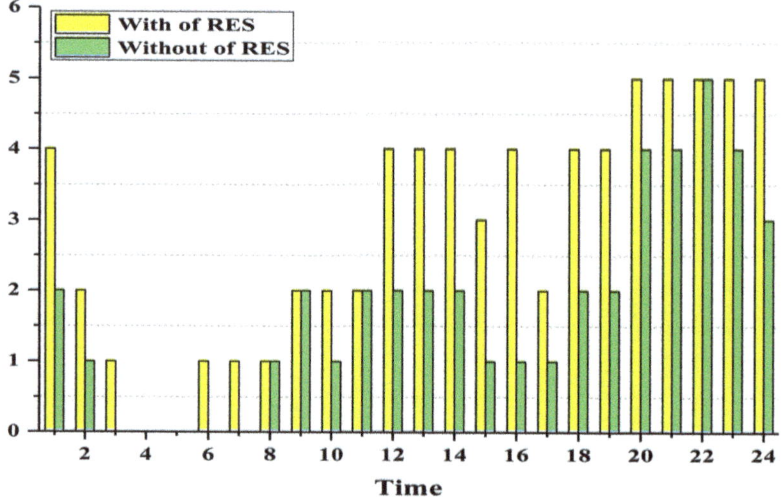

Fig. 3.32 Number of overloaded branch components

From the perspective of lines loading range, the number of overloaded lines before the presence of RES is equal to 70 during a day, but after adding RES, only 42 lines are operated in overload conditions. The number of overloaded branches for different intervals is shown in Fig. 3.32. As can be seen from Table 3.2, in the presence of RES, line loading in branches 1–2 and 2–3 were decreased by 13.41 and

Table 3.2 Lines loading ranges in presence/without RES

System's lines	Max. loading (%)		Min. loading (%)	
	Without RES	In presence RES	Without RES	In presence RES
01_02	**230.64**	**199.71**	**98.91**	**84.02**
02_03	**166.74**	**157.56**	**78.92**	**64.08**
03_04	139.20	129.38	60.81	46.18
04_05	125.72	115.35	56.84	42.06
05_06	115.11	104.12	53.52	38.55
06_07	58.68	62.68	30.76	20.45
06_26	57.68	57.03	19.69	14.15
02_19	56.80	54.05	14.12	15.36
07_08	50.75	46.29	27.69	10.38
26_27	50.13	43.42	17.50	12.60
08_09	43.59	40.89	24.74	10.04
27_28	41.65	40.56	14.99	12.34
03_23	39.76	39.46	9.92	10.03
19_20	39.30	39.33	10.01	7.79
09_10	38.28	35.62	22.36	8.31
10_11	33.67	33.22	20.09	9.46
28_29	33.00	31.02	12.23	7.17
11_12	29.18	30.82	17.83	4.38
23_24	27.97	28.77	7.13	7.21
29_30	25.96	26.24	9.83	15.41
20_21	25.32	25.42	6.49	6.51
12_13	24.90	24.83	15.61	6.74
13_14	20.04	21.09	12.80	10.25
30_31	18.11	16.22	6.75	5.14
14_15	15.43	15.85	10.03	4.26
24_25	14.75	15.64	3.78	5.38
21_22	13.54	15.17	3.49	3.82
31_32	12.61	15.05	4.72	5.34
15_16	11.05	13.59	7.31	3.50
16_17	7.38	11.60	4.95	0.56
32_33	7.18	8.90	2.69	3.06

5.505%, respectively, in compare with the initial status. The overall results for lines loading range are presented in Table 3.2.

In terms of busbars' voltage levels, the worst conditions have occurred on busbars 17 and 18. By connecting RES to the test system, the status of these busbars has improved somewhat. The complete results for voltage levels of different busbars are presented in Table 3.3.

Table 3.3 Voltage levels of each busbar in presence /without RES

	Max. voltage (pu)		Min. voltage (pu)	
System's busbars	Without RES	In presence RES	Without RES	In presence RES
03	0.93	0.94	0.85	0.86
04	0.89	0.92	0.77	0.79
05	0.86	0.90	0.70	0.72
06	0.79	0.85	0.54	0.57
07	0.78	0.85	0.52	0.59
08	0.75	0.83	0.47	0.51
09	0.71	0.81	0.40	0.45
10	0.68	0.75	0.34	0.37
11	0.67	0.74	0.33	0.36
12	0.66	0.74	0.31	0.34
13	0.63	0.71	0.26	0.28
14	0.62	0.70	0.24	0.26
15	0.61	0.70	0.23	0.25
16	0.60	0.70	0.22	0.24
17	**0.59**	**0.68**	**0.21**	**0.22**
18	**0.59**	**0.67**	**0.20**	**0.22**
19	0.98	0.99	0.96	0.96
20	0.96	0.96	0.87	0.88
21	0.96	0.96	0.86	0.86
22	0.95	0.95	0.84	0.85
23	0.92	0.94	0.82	0.85
24	0.91	0.93	0.79	0.82
25	0.90	0.92	0.77	0.81
26	0.78	0.85	0.53	0.58
27	0.78	0.84	0.51	0.56
28	0.75	0.82	0.45	0.48
29	0.73	0.81	0.42	0.47
30	0.72	0.81	0.40	0.46
31	0.71	0.81	0.38	0.45
32	0.71	0.81	0.37	0.50
33	0.71	0.77	0.37	0.43

References

1. Renewables energy global status report (2019) REN21. https://www.ren21.net/wp-content/uploads/2019/05/gsr_2019_full_report_en.pdf. Accessed 2019
2. Yong JHC, Wong J, Lim YS et al (2018) Assessment on various allocations of energy storages systems on radial distribution network for maximum PV systems penetration. Paper presented at the International Conference on Smart Grid and Clean Energy Technologies (ICSGCE), Kajang

3. Lammert G, Ospina LDP, Pourbeik P et al (2016) Implementation and validation of WECC generic photovoltaic system models in DIgSILENT PowerFactory. Paper presented at the IEEE Power and Energy Society General Meeting (PESGM), Location: Boston, MA
4. Jayawardena AV, Meegahapola LG, Perera S et al (2012) Dynamic characteristics of a hybrid microgrid with inverter and non-inverter interfaced renewable energy sources: A case study. Paper presented at the IEEE International Conference on Power System Technology (POWERCON), Auckland
5. Abu Dyak A, Alsafasfeh Q, Harb A (2018) Stability analysis of connected large-scale renewable energy sources into Jordanian national power grid. Int J Ambient Energy:2162–8246. https://doi.org/10.1080/01430750.2018.1501742
6. Bayindir R, Demirbas S, Irmak E et al (2016) Effects of renewable energy sources on the power system. Paper presented at the IEEE International Power Electronics and Motion Control Conference (PEMC), Varna
7. Hoseinzadeh B, Da Silva FF, Leth Bak C (2015) Active power deficit estimation in presence of renewable energy sources. Paper presented at the IEEE Power and Energy Society General Meeting, Denver, CO
8. Malkowski R, Bućko P, Jaskólski M et al (2018) Simulation of the dynamics of renewable energy sources with energy storage systems. Paper presented at the 15th International Conference on the European Energy Market (EEM), Lodz
9. Alavi O, Mohammadi K, Mostafaeipour A (2016) Evaluating the suitability of wind speed probability distribution models: a case of study of east and southeast parts of Iran. Energy Convers Manag 119:101–108
10. Ozay C, Celiktas MS (2016) Statistical analysis of wind speed using two-parameter Weibull distribution in Alaçatı region. Energy Convers Manag 121:49–54
11. Wind Turbines—Part (2005) 12-1: Power performance measurements of electricity producing wind turbines; international electrotechnical commission (IEC), IEC 61400-12-1. Geneva
12. Catalao JPS, Pousinho HMI, Mendes VMF (2011) Hybrid wavelet-PSO-ANFIS approach for short-term wind power forecasting in Portugal. IEEE Trans Sustain Energy 2(1):50–59
13. Galvani S, Hagh MT, Sharifian MBB et al (2018) Multiobjective predictability-based optimal placement and parameters setting of UPFC in wind power included power systems. IEEE Trans Ind Inf 15(2):878–888
14. Rastegar M, Fotuhi-Firuzabad M (2015) Load management in a residential energy hub with renewable distributed energy resources. Energ Buildings 107:234–242
15. Tavakoli Ghazi Jahani MA, Nazarian P, Safari A et al (2019) Multi-objective optimization model for optimal reconfiguration of distribution networks with demand response services. Sustain Cities Soc 47:101–114
16. Ahmadian A, Sedghi M, Fgaier H et al (2019) PEVs data mining based on factor analysis method for energy storage and DG planning in active distribution network: introducing S2S effect. Energy 175:265–277
17. Abapour S, Zare K, Mohammadi-Ivatloo B (2015) Dynamic planning of distributed generation units in active distribution network. IET Gener Transm Distrib 9(12):1455–1463
18. Nazari-Heris M, Madadi S, Hajiabbas MP et al (2018) Optimal distributed generation allocation using quantum inspired particle swarm optimization. In: Quantum computing: An environment for intelligent large scale real application. pp 419–432
19. Sangwongwanich A, Yang Y, Blaabjerg F (2016) High-performance constant power generation in grid-connectedPV systems. IEEE Trans Power Electron 31(3):1822–1825
20. Hung DQ, Mithulananthan N, Lee KY (2014) Determining PV penetration for distribution systems with time-varying load models. IEEE Trans Power Syst 29(6):3048–3058

Chapter 4
Power Quality and Harmonics Analysis in the Presence of Renewable Energy Sources

This chapter explains how to determine power quality indices by DIgSILENT PowerFactory in the presence of renewable energy sources. In PowerFactory software, *Harmonic Load Flow Analysis* and *Impedance Frequency Analysis* can be used to analyze power grids in terms of power quality. The outputs of these analyses show various indicators of the system's power quality especially harmonic distortion (HD) and total harmonic distortion (THD).

4.1 Power Quality

Power quality analysis in power grids can be fulfilled from a variety of perspectives, such as variations and unexpected events. Unbalanced loads and various harmonic sources are disturbances that lead to continuous variations on every cycle of the power system. However, unexpected events such as voltage dips are discrete in nature and may occur within a few cycles, and then may not repeat for several hours or days [1–3]. However, the voltage and current harmonics are the most important aspects of power quality [4]. From the power system operators' point of view, various types of harmonic sources in the power network can increase the temperature of rotating machines, increase the temperature of transformers and cables, reduce the lifetime of the various equipment, destroy smart meters, decrease system's reliability, and increase power losses [5, 6]. According to statistics released by the electric power research institute (EPRI), in recent years, commercial and industrial businesses in the United States have lost $45 billion per year due to power interruptions, out of which nearly $20 billion in losses were attributed to power quality problems [7]. This report clearly highlighted the importance of power quality issues in smart grids.

M. Zare Oskouei, B. Mohammadi-Ivatloo, *Integration of Renewable Energy Sources Into the Power Grid Through PowerFactory*, Power Systems,
https://doi.org/10.1007/978-3-030-44376-4_4

4.1.1 Power Quality Indices

According to conducted studies, many indicators have been introduced to evaluate power quality in power grids. Hence, the following indices can be used to evaluate the impact of renewable energy sources (RES) on the power quality performance of power grids:

I. **Harmonic distortion and total harmonics distortion**

The main cause of harmonic distortion in the power grid is related to the power electronic devices. These harmonic generating sources can cause voltage distortion in the power grid and especially near industrial consumers. The use of renewable sources in the vicinity of nonlinear loads is one of the most important factors in enhancing harmonic distortions [8]. The Fourier series concept is a very efficient tool for harmonic analysis. According to IEEE-519 standard [9], the harmonic distortion (HD) and the total harmonic distortion (THD) should be calculated at the point of common coupling (PCC) and must be less than the specified standard value. These indices for the system's current and voltage can be calculated by using Eqs. (4.1a, 4.1b, 4.1c, 4.1d, 4.1e, and 4.1f).

$$\mathrm{HD}_I = \frac{I_h}{I_1} \times 100 \tag{4.1a}$$

$$\mathrm{HD}_V = \frac{V_h}{V_1} \times 100 \tag{4.1b}$$

$$\mathrm{THD}_I = \frac{\sqrt{\sum_{h=2}^{\infty} \left| I_h \right|^2}}{I_1} \times 100 \tag{4.1c}$$

$$\mathrm{THD}_V = \frac{\sqrt{\sum_{h=2}^{\infty} \left| V_h \right|^2}}{V_1} \times 100 \tag{4.1d}$$

$$I_{rms}^2 = I_1^2 \times \left(1 + \mathrm{THD}_I^2\right) \tag{4.1e}$$

$$V_{rms}^2 = V_1^2 \times \left(1 + \mathrm{THD}_V^2\right) \tag{4.1f}$$

V_h and I_h denote the *h*th harmonic voltage and current measured at the PCC, respectively. V_1 and I_1 denote fundamental *rms* voltage and current values. Based on IEEE-519 standard, the maximum allowable total harmonic voltage and current distortions (THD_V, THD_I) must be as per the values of Tables 4.1 and 4.2.

Table 4.1 Voltage distortion limits [9]

Bus voltage at PCC	Individual harmonic (%)	Total harmonic distortions (%)
$V \leq 1$ kV	5	8
1 kV $\leq V \leq$ 69 kV	3	5
69 kV $\leq V \leq$ 161 kV	1.5	2.5
161 kV $\leq V$	1	1.5

Table 4.2 Current distortion limits [9]

SCR = I_{SC}/I_L	$3 \leq h < 11$	$11 \leq h < 17$	$17 \leq h < 23$	$23 \leq h < 35$	$35 \leq h < 50$
SCR < 20	4	2	1.5	0.6	0.3
20 ≤ SCR < 50	7	3.5	2.5	1	0.5
50 ≤ SCR < 100	10	4.5	4	1.5	0.7
100 ≤ SCR < 1000	12	5.5	5	2	1
1000 ≤ SCR	15	7	6	2.5	1.4

In Table 4.2, I_{sc} is the maximum short-circuit current at PCC, I_L is the maximum demand load current at the PCC under normal load operating conditions, and SCR is the short circuit ratio.

II. **Power factor**

The power factor is defined as the ratio of active power to the apparent power flowing in the circuit. The low power factor at the PCC can affect the energy-transfer efficiency of the system. Capacitor banks are used to improve the power factor range in power systems. According to existing standards, the acceptable range for the power factor is from 0.9 to 1.

III. **Voltage sag**

Unscheduled interchange among the renewable sources and power grid can cause voltage sags. The proposed approach for calculating the "sag score (*SS*)" by Detroit Edison is one of the most efficient methods in this field [10]. This method is based on the calculation of the voltage magnitudes of the three phases at the PCC. According to this method, the *SS* can be calculated by Eq. (4.2).

$$SS = 1 - \left(\frac{V_A + V_B + V_C}{3} \right) \tag{4.2}$$

where V_A, V_B, and V_C are the voltage magnitude of the three-phase power system.

IV. **Power frequency deviation**

Random variations in the generation of RES cause frequency deviation in the power system because of unscheduled and intermittent power flow in transmission lines. According to EN 50160 standard, the allowable range of frequency changes during the period of operation shall be ±4% of the rated frequency (50 or 60 Hz) [11]. The power frequency deviation is calculated using Eq. (4.3).

$$\text{FDR} = \frac{|f_m - f_r|}{f_r} \times 100 \tag{4.3}$$

where f_m is the measured fundamental frequency at the PCC and f_r is the rated system frequency.

4.2 Harmonic Sources Model in PowerFactory

Before applying *Harmonic Load Flow Analysis* in PowerFactory, the harmonic model for all components of the network must be determined. The nonlinear loads (*ElmLod*), thyristor rectifiers (*ElmRec*), PWM-converters (*ElmVsc*, *ElmVscmono*), static generators (*ElmGenstat*), voltage sources (*ElmVac*), and current sources (*ElmIac*) are recognized as the harmonic generating sources in power systems. The use of power electronic devices in different forms will cause the current harmonic distortion in the grid, and then these harmonics will generate voltage harmonics. Therefore, users can obtain a precise model of the power grid from *Harmonic Load Flow Analysis* only by determining the harmonic current sources.

4.2.1 Assignment of Harmonic Current Source to the General Loads

Approximately 54% of power quality problems are related to the general loads [12]. The following steps should be taken into account for harmonic modeling of the general loads in PowerFactory:

- *Double-clicking on the desired general load*;
- *Go to the "Basic Data" tab and select "New Project Type-General Load Type" from "Type" section*;
- *Go to the "Harmonics/Power Quality" tab and choose the "Current Source" model using the drop-down list on the "Load Model" box.*

As discussed earlier, we can model general loads in different types such as balanced three-phase, unbalanced three-phase, two-phase, and single-phase with different configurations in DIgSILENT PowerFactory. When using balanced three-phase loads, even-order harmonics (h = 2, 4, 6, 8, …) and also multiples of third harmonics (h = 3, 6, 9, …) will be eliminated by using the combined star-delta connection. Hence, even order and multiples of third harmonics will not be applied in the analysis of balanced three-phase loads. After performing the mentioned steps, the following steps should be done to allocate the harmonic current source model to the general loads. These steps are shown in Fig. 4.1:

- *Go to the "Harmonics/Power Quality" tab through the setting dialogue* (marked ❶ in Fig. 4.1);
- *Select "New Project Type-Harmonic Currents" from "Harmonic Current Injections" box* (marked ❷ in Fig. 4.1);
- *There are three types of harmonic models for general loads (1) Balanced, (2) Unbalanced, and (3) IEC 61000. Users can choose one of three available harmonic models according to their requirements* (marked ❸ in Fig. 4.1);
- *Right-click on the "Harmonics" box then select "Append n Rows." After that allocate the number of needed rows. As can be seen in the balanced mode, even-order and multiples of third harmonics are not created by PowerFactory* (marked ❹ in Fig. 4.1);
- *The ratio of the magnitude of each harmonic order to the fundamental harmonic must be defined in the first column. Also, the ratio of the phase of each harmonic order to the fundamental harmonic must be defined in the second column* (marked ❺ in Fig. 4.1).

In the unbalanced mode, the harmonic ratios must be measured for all three phases (A, B, and C). When measurements of the phase of each harmonic order are

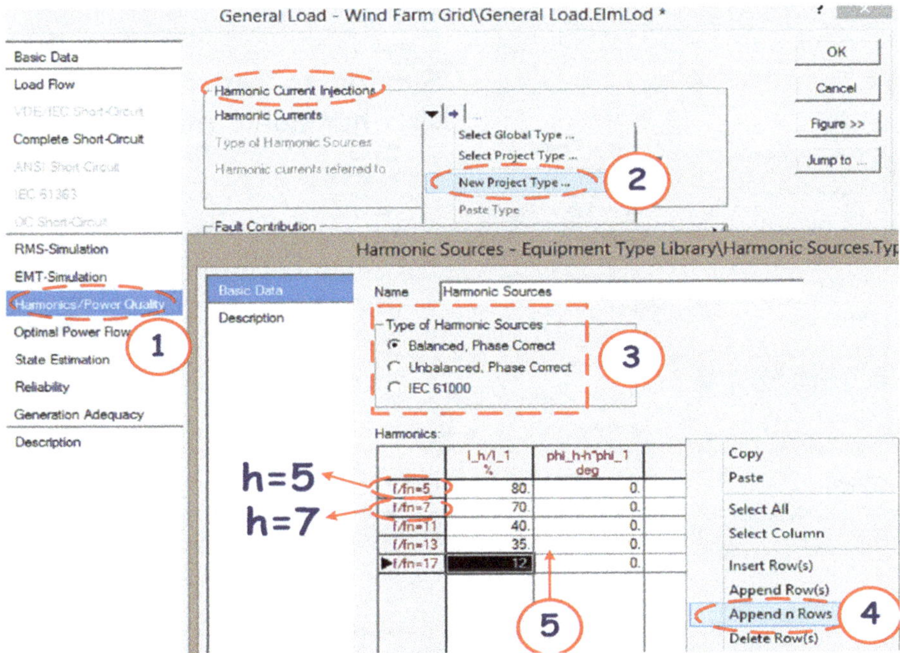

Fig. 4.1 Process of defining harmonic current source to the general loads

not available, the best solution is to use IEC 61000 mode. In addition to the integer harmonics, it is possible to use non-integer harmonics in IEC 61000 mode.

4.2.2 Assignment of Harmonic Model to RES

Users can investigate the role of RES in the injected harmonics into the power grid using two methods. The first method is the same as in the previous subsection, in which it is fully explained how to assign harmonic models to general loads. The second method is based on the use of the Norton equivalent circuit for RES according to the dynamic model of each system. Each Norton equivalent circuit is defined by an equivalent resistance and reactance. Due to the skin effect, resistance and reactance are usually frequency-dependent. In this method, the relation between resistance and reactance with frequency can be used to assign the harmonic model to renewable sources.

In DIgSILENT PowerFactory, the frequency dependence of Norton equivalent circuit elements can be defined in two ways. For this purpose, according to Fig. 4.2, users must first enter the renewable sources' setting page and then select one of the Norton equivalent circuit elements via the "Harmonics/Power Quality" tab. In the first method, users can define frequency characteristics for the Norton equivalent circuit elements through a polynomial function. The polynomial function used in

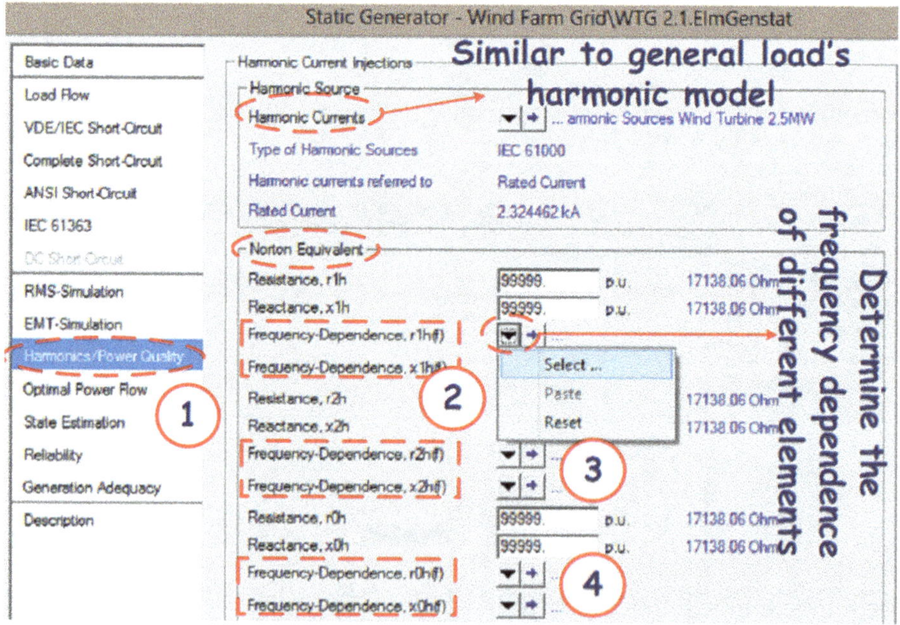

Fig. 4.2 Norton equivalent circuit elements of renewable sources

DIgSILENT is shown in Eq. (4.4). The parameters a and b should be specified based on the inherent characteristics of different elements. The steps to access the "Frequency Polynomial Characteristic" dialogue are shown in Fig. 4.3.

$$R(h) = (1 - a) + a \cdot \left(\frac{f_h}{f_1} \right)^b \tag{4.4}$$

where f_1 is the fundamental frequency and f_h is the variable frequencies.

In the second method, users can define frequency characteristics for the Norton equivalent circuit elements through a frequency table. The required steps to apply the frequency table for the various elements include:

- *Choose "Parameter Characteristic-Vector" through the "New Object" button* (marked ❷ in Fig. 4.3);
- *Press the "Select" button from "Scale" tab and then choose "Frequency Scale" using the drop-down list* (marked ❶, ❷ in Fig. 4.4a);
- *In the pop-up window, enter the desired "Scale Values" for system frequency using "Append n Rows" option* (marked ❶ in Fig. 4.4b);
- *Then go back to the "Parameter Characteristic-Vector" window and determine how the elements depend on each frequency through the created table* (marked ❷ in Fig. 4.4b).

These steps are shown in Fig. 4.4a and b.

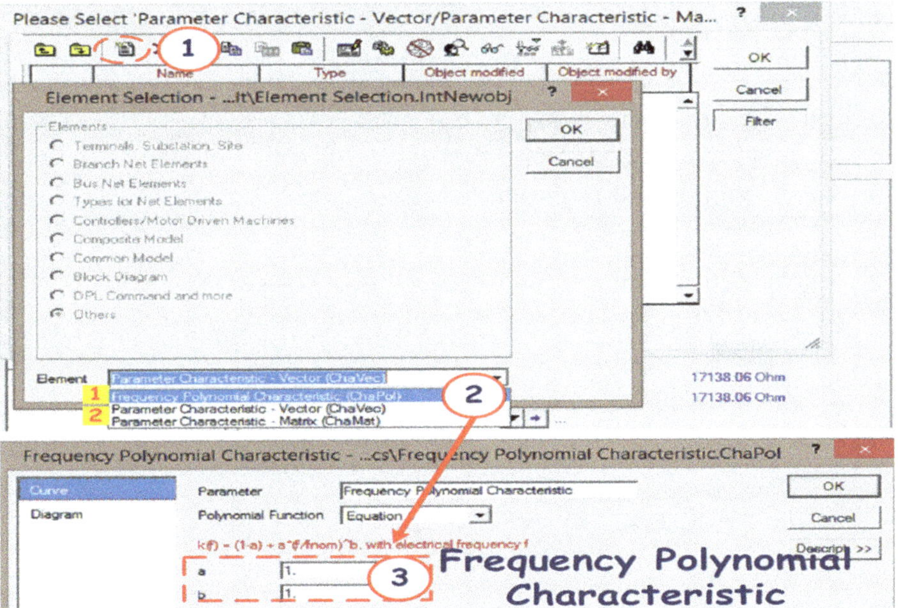

Fig. 4.3 Process of defining frequency polynomial characteristic to RES

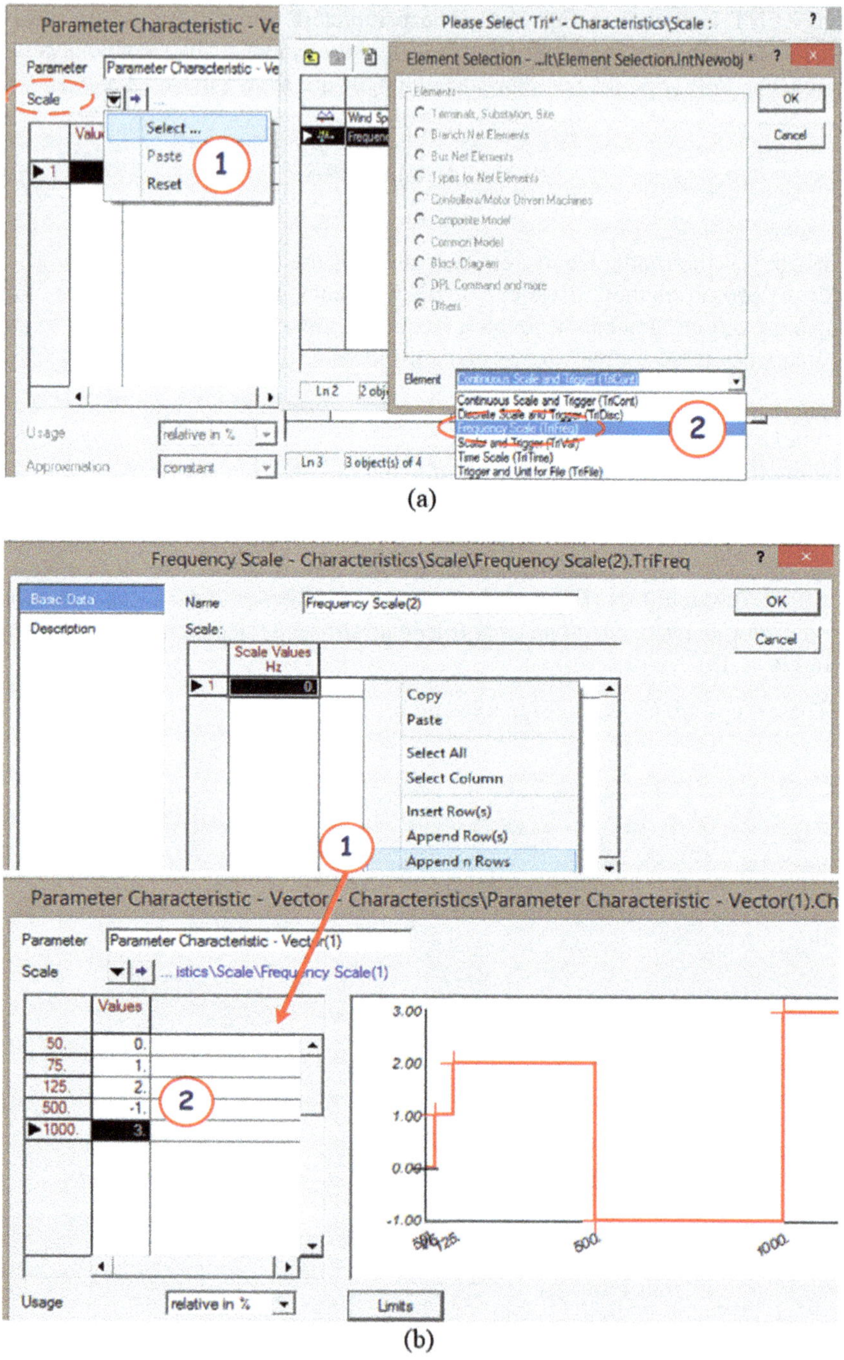

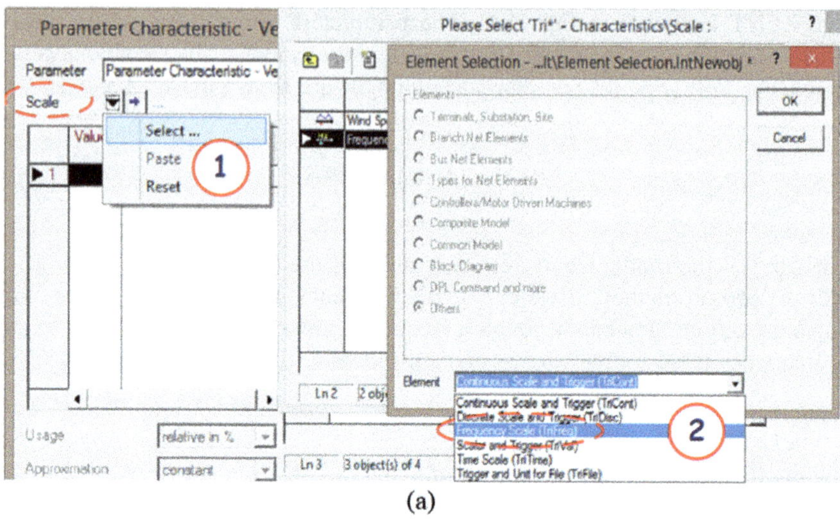

(a)

(b)

Fig. 4.4 Process of determining frequency table for the Norton equivalent circuit elements (**a**) creating frequency table and (**b**) assigning desired values

4.2.3 *Assignment of Harmonic Model to Other Elements*

Other important elements that generate harmonic distortions in power systems are:

- **External grid and AC voltage source**: These sources cause voltage harmonic distortion in the power grid. In fact, these sources play the role of background harmonic for the downstream system.
- **AC current source**
- **Lines and cables**
- **Inverter, rectifier, and PWM converter**

It should be noted that the presented methods in the previous sub-sections can be used to assign the current and voltage harmonic model to all these elements.

4.3 Harmonics/Power Quality Analysis

Harmonic Load Flow Analysis is the most important tool available in the PowerFactory to assess various power quality indices in power grids. Harmonic load flow tool separately performs load flow calculation at different harmonic orders and displays their results to the user. This tool is also able to investigate the impact of short and long term flickers due to continuous operation and switching operations of renewable sources and power electronic devices on power system performance. To perform the harmonic load flow simulation, press the "Calculate Harmonic Load Flow" command located on the "Calculation-Harmonics/Power Quality" section. Users can define various conditions to apply *Harmonic Load Flow Analysis*, which are described below.

I) **Basic Options**

- *Network representation*: The network under study can be evaluated based on the used elements in both balanced and unbalanced modes. By using the balanced mode, characteristic harmonics appear in the positive or negative sequence component (5th, 7th, 13th, 19th, etc.). But in the unbalanced mode, it is possible to analyze third-order harmonics, even-order harmonics, and inter-harmonics;
- *Calculate harmonic load flow*: Power quality analysis can be performed for only one frequency (output frequency) or all harmonic frequencies. It should be noted that if the "Frequency Single" option is selected, the calculations will be performed based on "Harmonic Order" and "Nominal Frequency" values.

$$h = \frac{f_h}{f_1}$$

where h is the harmonic order, f_h is the output frequency, and f_1 is the nominal frequency (50 or 60 Hz). After performing the harmonic load flow, the results shown in the result boxes are based on the specified "Harmonic Order" value.

- Ability to calculate flicker and short circuit power at fundamental frequency.
- Ability to change the initial conditions of *Load Flow Analysis*.

II) **IEC 61000-3-6**

This tab applies if users have used standard 61,000 in defining harmonic sources. This standard uses the "second summation law" for the *Harmonic Load Flow Analysis* of the system. The required parameters to use this standard are available in the software by default.

III) **Advanced Options**

PowerFactory can calculate the HD and THD indices based on the IEEE and DIN standards. The difference between these two standards is in the fundamental current rate. The equations used to determine the fundamental current by these two standards are presented in Eqs. (4.5a and 4.5b).

$$I_{1,\mathrm{IEEE}} = |I_1| \tag{4.5a}$$

$$I_{1,\mathrm{DIN}} = \sqrt{\sum_{h=1}^{n} I_h^2} \tag{4.5b}$$

4.3.1 Display of HD and THD Indices

There are two ways to display THD and HD indices in different busbars of the power grid:

First way:

- *Hold down* ***<u>Ctrl</u>*** *and select the desired busbars to display THD and HD* indices;
- **To display HD index:** *Right-click on the selected busbars then choose "Harmonic Distortion" from "Show-Distortion Diagram" option;*
- **To display THD index**: *Right-click on the selected busbars then choose "Total Harmonic Distortion" from the "Show-Bar Diagram" option.*

Second way:

- *Right-clicking on the desired busbars*;
- *Select "Results for Harmonic Load Flow" option from the "Define" section* (marked ❶, ❷ in Fig. 4.5);
- *In the pop-up window, double-clicking on the created elements* (marked ❸ in Fig. 4.5);

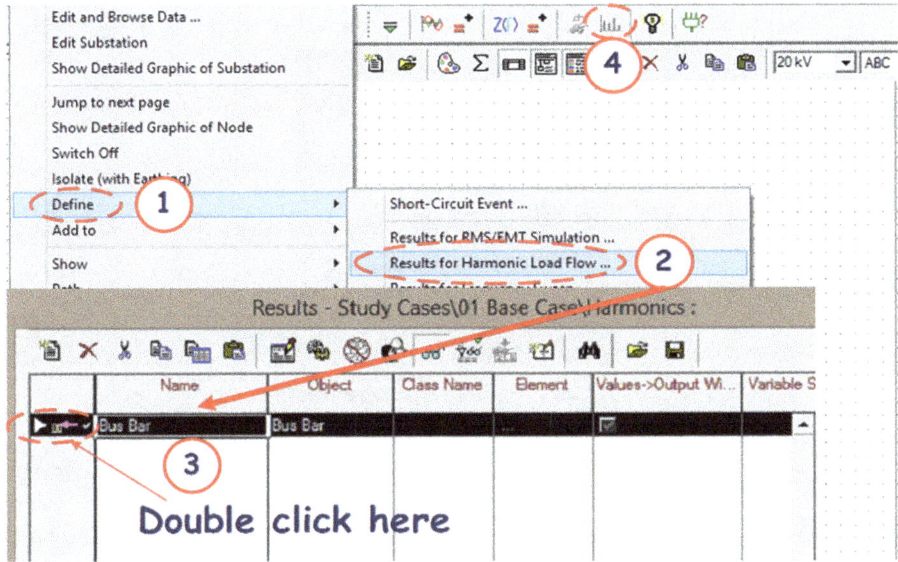

Fig. 4.5 Process of displaying HD and THD indices

- *In the "Variable Selection Monitor" window, go to the "Harmonic/Power Quality" tab;*
- *Go to the "Currents, Voltages, and Powers" section using the drop-down list on the "Variable Set" box;*
- *Choose THD and HD indices from the "Available Variables" box;*
- *Press the "Create Distortion Plot" button and then choose the desired elements and variables to display* (marked ❹ in Fig. 4.5).

4.3.2 Display Waveform of Busbars' Voltage and Lines' Current

Users can draw the waveforms of the current flowing through lines and the voltage of busbars in the presence of different harmonic sources using the "Waveform" tool. After performing the *Harmonic Load Flow Analysis*, the software transmits the results of the harmonic load flow into real time by using Fourier transform. The required steps to draw the waveforms of the current and the voltage in the network under study are as follows:

- *Right-clicking on the desired elements*;
- *Select "Results for Harmonic Load Flow" option from the "Define" section;*
- *In the pop-up window, double-clicking on the created elements*;

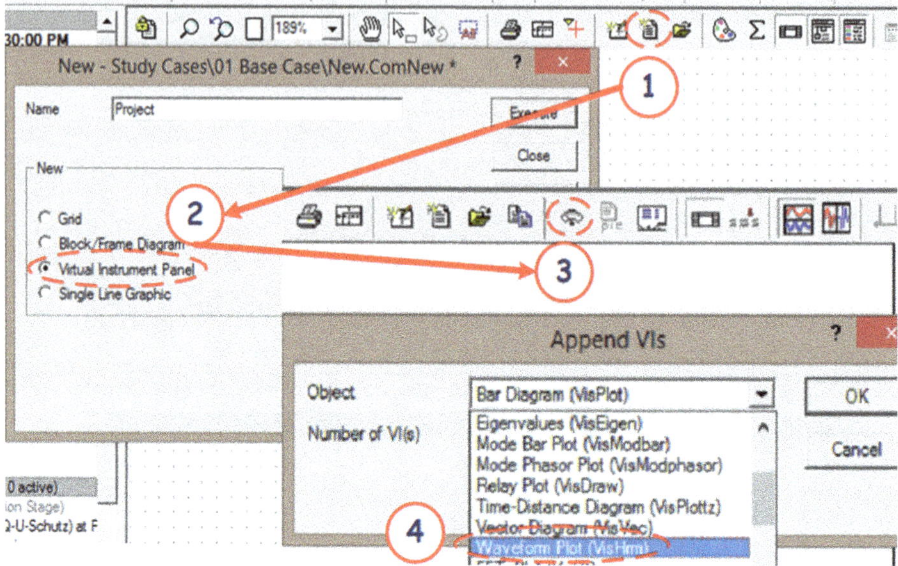

Fig. 4.6 Process of plotting waveforms of the system's current and voltage

- *In the "Variable Selection Monitor" window, go to the "Harmonic/Power Quality" tab;*
- *Go to the "Currents, Voltages, and Powers" section using the drop-down list on the "Variable Set" box;*
- *For lines, the* ***current magnitude*** *with the* ***current angle,*** *and for busbars, the* ***voltage magnitude*** *with the* ***voltage angle*** *must be selected*;
- *Press the "Insert New Graphic" button located on the main menu and then select "Virtual Instrument Panel" option* (marked ❶, ❷ in Fig. 4.6);
- *In the opened window, create a new "Waveform Plot"* via *the "Append New VI(S)" button* (marked ❸, ❹ in Fig. 4.6);
- *Double-clicking on the created window and then choose the desired elements with their* ***magnitude*** *and* ***angle***.

These steps are shown in Fig. 4.6.

4.3.3 Impedance Frequency Analysis

The network operator can detect the resonant frequency of the power grid using *Impedance Frequency Analysis*. Therefore, the purpose of this analysis is to determine the harmonic order that leads to the resonant frequency. The network operator can prevent resonant phenomena in the network by adding appropriate filters after detecting the resonant frequency. In PowerFactory, *Impedance Frequency Analysis* can be performed by considering the effect of either self-impedances or mutual impedances.

4.3.3.1 Effect of Self-Impedance

In this part of the *Impedance Frequency Analysis*, the equivalent impedance of the network is obtained from the PCC point of view. The following steps must be taken to this end:

- *Right-clicking on the desired busbar (PCC busbar)*;
- *Select "Results for Frequency Sweep" option from the "Define" section;*
- *In the "Variable Selection Monitor" window, go to the "Harmonic/Power Quality" tab;*
- *Go to the "Currents, Voltages, and Powers" section using the drop-down list on the "Variable Set" box;*
- *Choose the magnitude and angle of impedance from the "Available Variables" box*;
- *Press the "Calculate Impedance Frequency Characteristic" button located on the "Calculation-Harmonics/Power Quality" section;*
- *In the setting window, "Start Frequency," "Stop Frequency," and "Step Size" should be determined to perform frequency sweep analysis. Please note that the computing speed will increase by enabling the "Automatic Step Size Adaptation" option.*

View the results of the analysis:

- *Press the "Insert New Graphic" button located on the main menu and then select "Virtual Instrument Panel" option;*
- *In the opened window, create a new "Bar Diagram"* via *the "Append New VI(S)" button*;
- *Double-clicking on the created window and then choose the desired busbar as the element and magnitude and angle of impedance as variables.*

Tips on adjusting X-axis to better display outputs:

- *Double-clicking on the graphical window and then go to the "X-Axis" tab* (marked ❶ in Fig. 4.7);
- *Select "Page" option from the "Scale" box* (marked ❷ in Fig. 4.7);
- *Go to the "Used Axis" menu and determine "Frequency" or "Harmonic Order" as the "X-Axis Variable"* (marked ❸, ❹ in Fig. 4.7);
- *Go to the "Advanced" tab and select "Curves" from the "Presentation" box.*

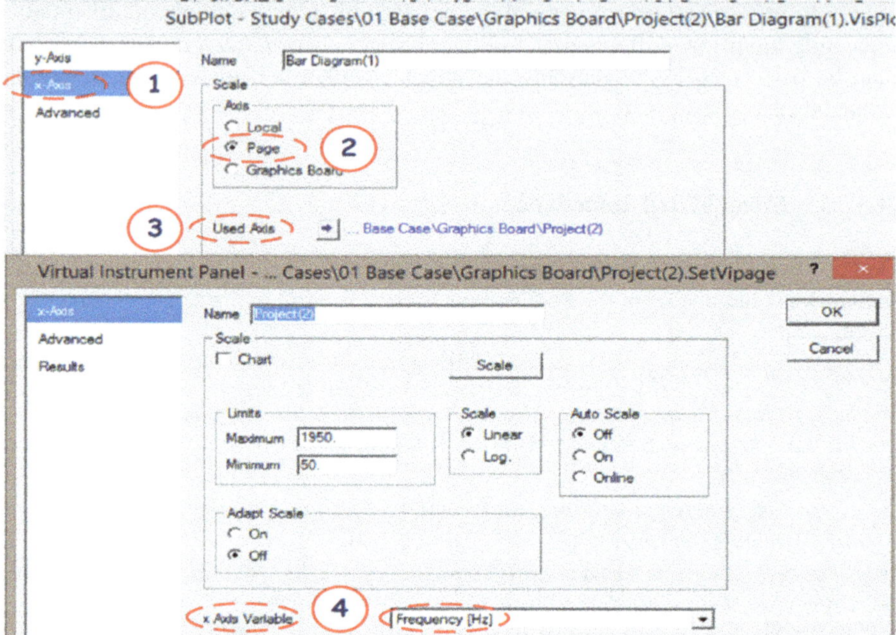

Fig. 4.7 Adjusting X-axis to display outputs

4.3.3.2 Effect of Mutual Impedance

This analysis is used to obtain the frequency spectrum of the mutual impedances between different equipment. Suppose two wind farms, as in Fig. 4.8, are several kilometers apart. The main purpose of this part of the simulation is to investigate the effect of virtual coupling between two neighbor busbars on the resonant frequency. The following steps need to be taken to define a virtual coupling between two busbars:

- *Go to the active project's library from the "Data Manager" window;*
- *Go to the "Grid" section located on the "Network Model-Network Data" section* (marked ❶ in Fig. 4.9);
- *Select desired busbars and then right-click on the selected busbars then choose "Mutual Data" option from "Define" section* (marked ❷–❹ in Fig. 4.9);
- *In the pop-up window, double-clicking on the created object;*
- *In the "Variable Selection Monitor" window, go to the "Harmonic/Power Quality" tab;*
- *Go to the "Calculation Parameter" section using the drop-down list on the "Variable Set" box;*

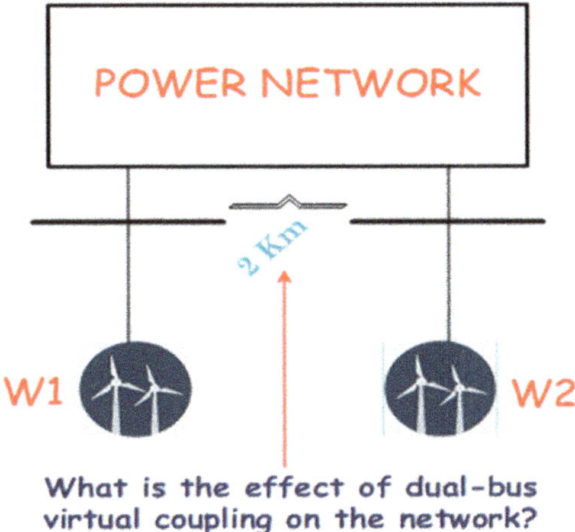

Fig. 4.8 Effect of dual-bus virtual coupling on the network

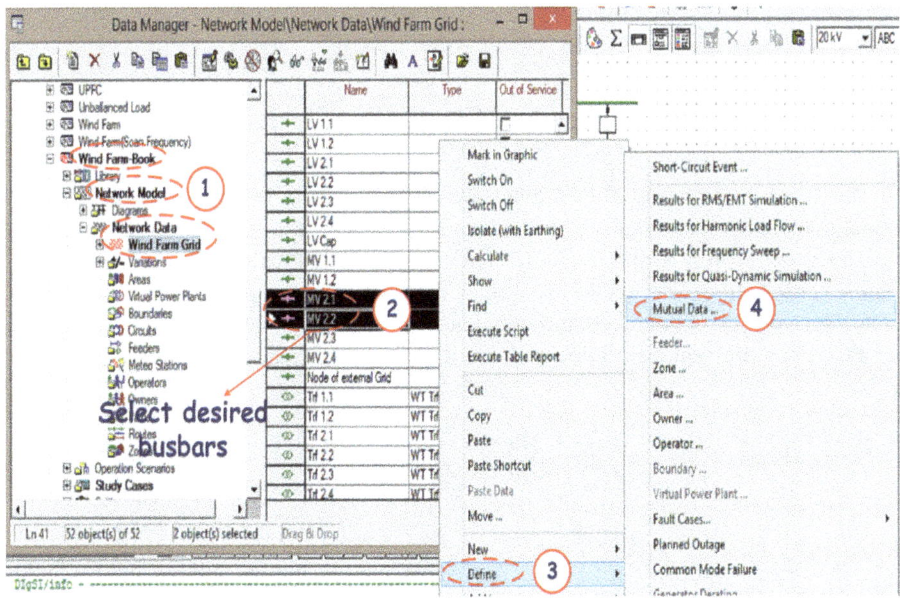

Fig. 4.9 Process of determining virtual coupling between two busbars

- *Choose the magnitude and angle of coupling impedance (Z_12 or Z_21) from the "Available Variables" box;*
- *Press the "Calculate Impedance Frequency Characteristic" button located on the "Calculation-Harmonics/Power Quality" section;*
- *In the setting window, "Start Frequency," "Stop Frequency," and "Step Size" should be determined to perform frequency sweep analysis.*

These steps are shown in Fig. 4.9. Finally, the results can be observed according to the described steps in Sect. 4.3.3.1.

4.4 Evaluation of Power Quality Indices in the Presence of RES

The test system used to evaluate various power quality indices in the presence of several wind turbines is shown in Fig. 4.10. This test system is a modified state of the existing example (Wind Farm) in PowerFactory 15.1. This system consists of six wind turbines and seven general loads. The test system is connected to the 20 kV grid via busbar 1 (B_1). This busbar acts as a PCC point between the upstream network and the test system. The branch data and various parameters for the test system are included in the software by default. The operation of wind turbines is based on the power-speed curve. The used power–speed curve can be seen in Fig. 4.11. The load data for the test system is given in Table 4.3.

4.4.1 Impact of Balanced and Unbalanced Modes

In this section, power quality indices are investigated in balanced and unbalanced modes. General loads and wind turbines as the harmonic generation sources were modeled according to the available information in Tables 4.4 and 4.5. *Harmonic Load Flow Analysis* is used to investigate the test system under different conditions. The THD values in balanced and unbalanced modes at different busbars, especially at PCC, are presented in Table 4.6. As it can be observed in Table 4.6, the THD

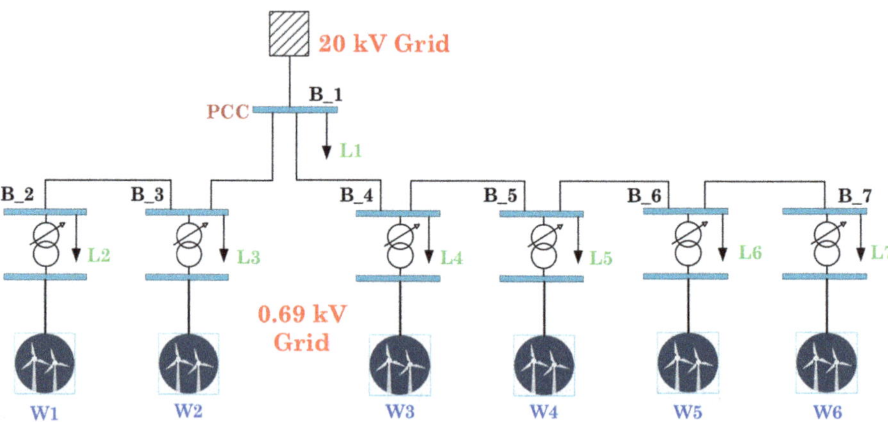

Fig. 4.10 Electrical network of the test system

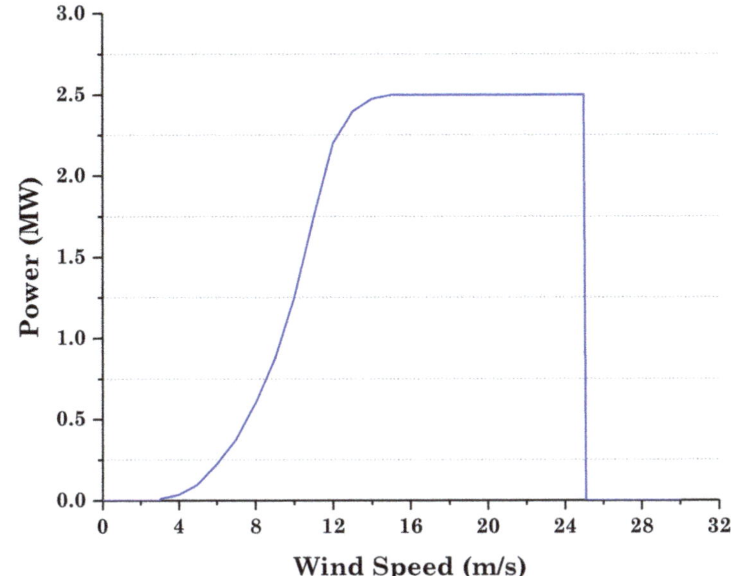

Fig. 4.11 Power-speed curve for wind turbines

Table 4.3 Load data for the test system

Load	Active power (MW)	Power factor
L1	4.5	0.8-lag
L2	1.5	0.85-lag
L3	1.7	0.83-lag
L4	2.2	0.9-lag
L5	1.8	0.78-lag
L6	2.4	0.9-lag
L7	2	0.88-lag

Table 4.4 Harmonic generation by wind farms and loads in balanced mode

Harmonic order	I_h/I_1 (%)	Phase (deg)
$h = 5$	20	180
$h = 7$	14.28	0
$h = 11$	9.09	180
$h = 13$	7.69	0
$h = 17$	5.88	180
$h = 19$	5.26	0
$h = 23$	4.34	180
$h = 25$	4	0

Table 4.5 Harmonic generation by wind farms and loads in unbalanced mode

Harmonic order	Ia_h/Ia_1 (%)	Ib_h/Ib_1 (%)	Ic_h/Ic_1 (%)	Phase_a (deg)	Phase_b (deg)	Phase_c (deg)
$h = 3$	33.33	33.33	33.33	180	180	180
$h = 5$	20	20	20	0	0	0
$h = 7$	14.28	14.28	14.28	180	180	180
$h = 9$	11.11	11.11	11.11	0	0	0
$h = 11$	9.09	9.09	9.09	180	180	180
$h = 13$	7.69	7.69	7.69	0	0	0
$h = 15$	6.66	6.66	6.66	180	180	180
$h = 17$	5.88	5.88	5.88	0	0	0
$h = 19$	5.26	5.26	5.26	180	180	180
$h = 21$	4.76	4.76	4.76	0	0	0
$h = 23$	4.34	4.34	4.34	180	180	180
$h = 25$	4	4	4	0	0	0
$h = 27$	3.71	3.71	3.71	180	180	180

Table 4.6 Comparison of THD in balanced and unbalanced mode

	Total Harmonic distortion (%)	
Busbars	Balanced mode	Unbalanced mode
B_1 (PCC)	**43.32**	**140.84**
B_2	44.85	144.69
B_3	44.62	144.04
B_4	48.28	160.32
B_5	49.56	164.72
B_6	50.48	167.65
B_7	51.07	169.15

value at PCC (B_1) has the lowest value compared to other busbars. As expected, the THD values in the unbalanced mode are much higher than the balanced mode.

The HD values for each harmonic order in balanced and unbalanced modes are shown in Figs. 4.12 and 4.13. In the balanced mode, the highest HD value in all busbars belongs to the fifth and seventh harmonic orders, which are approximately 25%. In the unbalanced mode, the highest HD value belongs to the 29th harmonic order. As can be seen from Fig. 4.13, the operation of the test system in unbalanced mode increases the HD values.

The waveform of the current flowing between the PCC and external grid (20 kV grid) in the balanced and the presence of harmonic sources is shown in Fig. 4.14. As can be seen from this figure, due to the low THD values, the deviation from the sinusoidal waveform is negligible. Also, the voltage waveform at the PCC (B_1) for phase *A* in unbalanced mode is shown in Fig. 4.15. Unlike the balanced mode, the voltage waveform deviation from the sinusoidal state is stronger under the unbalanced condition.

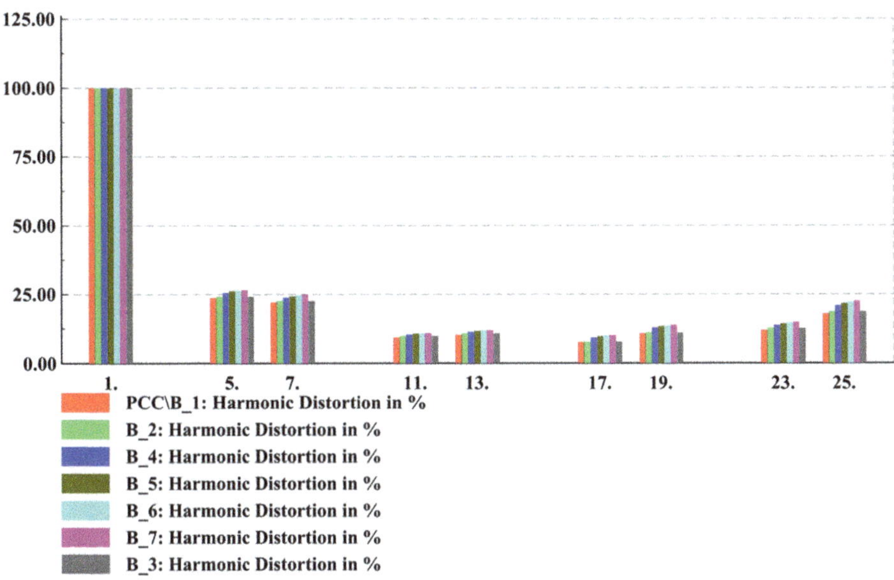

Fig. 4.12 Harmonic distortion in balanced mode

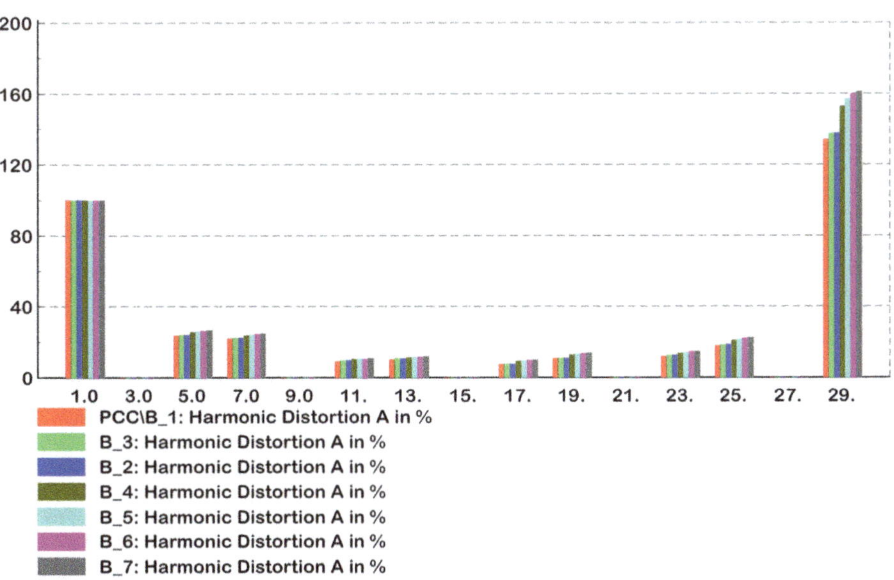

Fig. 4.13 Harmonic distortion in unbalanced mode

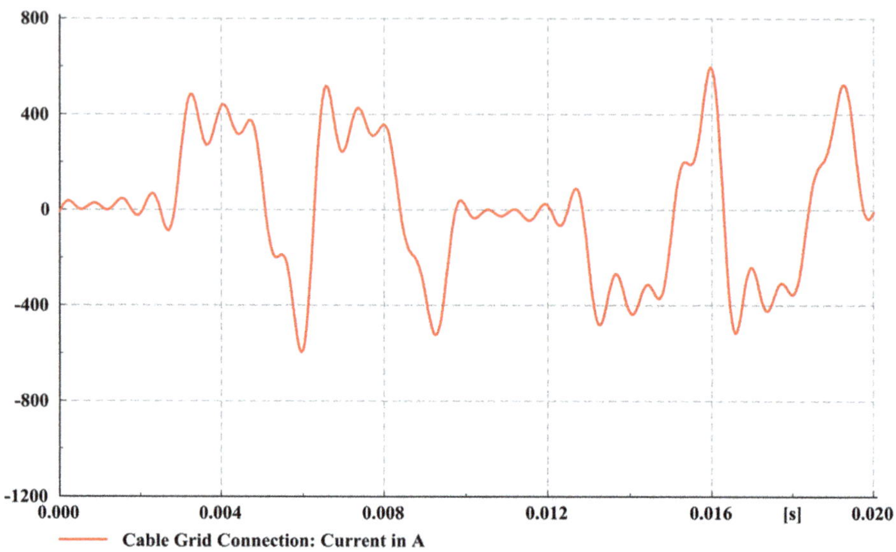

Fig. 4.14 Waveform of the current flowing between PCC and external grid in balanced mode

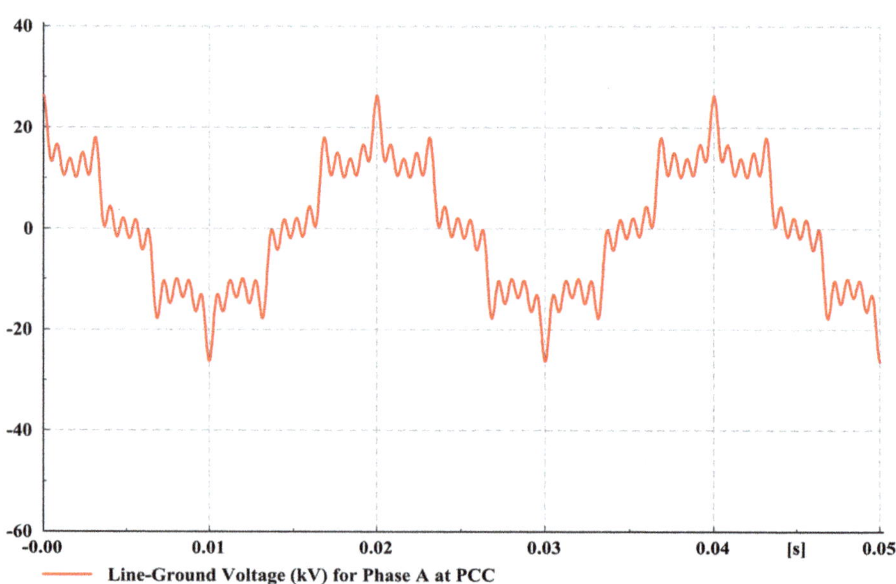

Fig. 4.15 The voltage waveform for phase *A* at the PCC in unbalanced mode

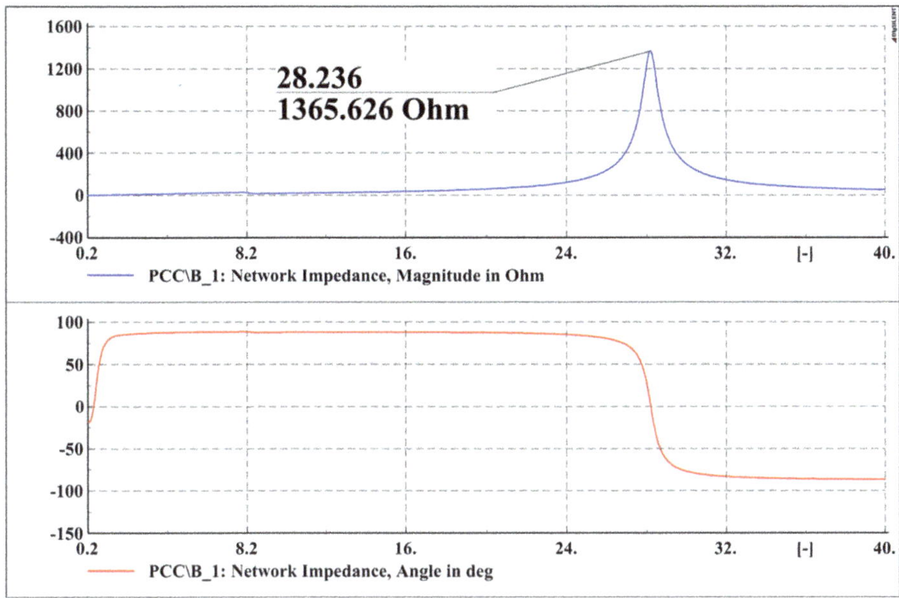

Fig. 4.16 The resonance impedance of the test system in balanced mode

4.4.2 *Reducing the Effects of Resonant Frequency*

In this section, the resonant frequency of the test system is evaluated *Impedance Frequency Analysis*. The resonance impedance and its angle in the balanced mode can be seen in Fig. 4.16. According to this figure, the resonance condition occurred at 28.236th harmonic. In this harmonic order, the magnitude of the resonance impedance is 1365.626 Ω. If the grid is not technically in good condition, the resonant frequency will occur at low harmonic orders. In this situation, the appropriate size of capacitive filters can be calculated using this analysis to reduce the effects of resonance phenomena.

References

1. Naderi Y, Hosseini SH, Mohammadi-Ivatloo B et al (2018) An overview of power quality enhancement techniques applied to distributed generation in electrical distribution networks. Renew Sust Energ Rev 93:201–214
2. Elbasuony GS, Abdel Aleem SHE, Ibrahim A et al (2018) A unified index for power quality evaluation in distributed generation systems. Energy 149:607–622
3. Montoya FG, García-Cruz A, Montoya AG et al (2016) Power quality techniques research worldwide: a review. Renew Sust Energ Rev 54:846–856

4. Naderi Y, Hosseini SH, Mohammadi-Ivatloo B et al (2019) An optimized direct control method applied to multilevel inverter for microgrid power quality enhancement. Int J Electr Power Energy Syst 107:496–506
5. Abdul Kadir AF, Mohamed A, Shareef H (2011) Harmonic impact of different distributed generation units on low voltage distribution system. Paper presented at the IEEE International Electric Machines & Drives Conference (IEMDC), Niagara Falls, ON
6. Bollen MHJ (2003) What is power quality? Electr Power Syst Res 66(1):5–14
7. Electric Power Research Institute (EPRI), The cost of power disturbances to industrial and digital economy companies, [Online]. Available at: https://www.epri.com/#/pages/product/3002000476/?lang=en-US/
8. Sakar S, Balci ME, Abdel Aleem SHE et al (2017) Increasing PV hosting capacity in distorted distribution systems using passive harmonic filtering. Electr Power Syst Res 148:74–86
9. IEEE Std 519-2014 (Revision IEEE Std 519-1992). Recommended practice and requirements for harmonic control in electric power systems. IEEE 2014:1–29
10. Polycarpou A (2011) Power quality and voltage sag indices in electrical power systems. In: Romero G (ed) Electrical generation and distribution systems and power quality disturbances. IntechOpen, Rijeka, pp 140–160
11. Markiewicz H, Klajn A (2004) Voltage disturbances standard EN 50160
12. Senra R, Boaventura WC, Mendes EM (2017) Assessment of the harmonic currents generated by single-phase nonlinear loads. Electr Power Syst Res 147:272–279

Chapter 5
Electrical Challenges Associated with Integrating Renewable Energy Sources into Power Grids

This chapter presents a comprehensive analysis of the effect of renewable energy sources (RES) in the technical parameters of the power system. In addition, several standard case studies presented to compare the impact of RES in various operation modes with conventional generating units and the battery energy storage system (BESS). It should be noted that the main focus of this chapter is to examine the electrical challenges resulting from the misuse of RES during the operation of power systems. Therefore, it is assumed that the optimal scheduling problem considering RES was performed on the desired test system by default. PowerFactory provides very useful packages for the system operator to investigate the status of the power grid in coordination with various renewable sources technologies, and perform the required technical analysis. Required operation tools to incorporate renewable sources in smart power grids include:

- Power flow analysis with low to high penetration of RES.
- Short-circuit studies and contingency analysis with stochastic models that capture the uncertainty of RES.
- Short-term rescheduling according to network performance after performing various simulations in the presence of RES as well as BESS.

5.1 Short-Circuit Analysis in the Presence of RES

Power grids can be divided into weak or strong categories in terms of "strength level" [1]. Improper use of renewable energy sources (RES) as inverter-based resources can have detrimental effects on weak electric systems. In this condition, common engineering challenges that system operators have had to face include: (1) transmission line overloading, (2) voltage deviation, and (3) low short circuit ratio (SCR) [2].

M. Zare Oskouei, B. Mohammadi-Ivatloo, *Integration of Renewable Energy Sources Into the Power Grid Through PowerFactory*, Power Systems,
https://doi.org/10.1007/978-3-030-44376-4_5

To analyze the performance of the power grid in the presence of RES, the strength level of the desired network at each bus must be determined. The strength level of a network is related to the sensitivity of the inverter-based source terminal voltage to its current injection changes [3]. This sensitivity in strong networks is less than in weak networks. Hence, in weak power networks, it is necessary to carry out rigorous technical studies to implement RES expansion plans [4]. The SCR index is an appropriate index to determine the relative strength of the power networks. This index is defined as the ratio of the short-circuit power level at the RES interconnection point to the nominal capacity of the RES. Equation (5.1) shows the mathematical formula of the SCR index [1].

$$SCR_i = \frac{SK_i''}{Pn_i} \tag{5.1}$$

where i is the candidate busbar, SK_i'' is the short-circuit power level at the ith busbar, Pn_i is the power injection of the RES.

According to the IEC 60909 standard [5], sufficient strength at the desired busbar is obtained when the SCR is less than 3. Also, the short-circuit power level is derived from the three-phase short-circuit current, which is given by Eq. (5.2).

$$SK_i'' = \sqrt{3} \times IK_i'' \times U_i^n \tag{5.2}$$

where U_i^n is the nominal line-to-line voltage at the ith busbar and IK_i'' is the three-phase short-circuit current at the ith busbar.

For the reasons stated earlier, system operators must calculate the short-circuit power level at the RES interconnection point to the network to examine the status of the network.

5.1.1 Basic Settings to Evaluate Short-Circuit Power Level

To perform *Short-Circuit Analysis* in the presence of RES and to achieve the expected results of the system operator, basic settings must first be applied in DIgSILENT PowerFactory. The required settings are divided into two categories: (1) general settings for *Short-Circuit Analysis* from the "*Short-Circuit Calculation*" toolbox and (2) local settings related to the structural specifications of each renewable source.

5.1.1.1 General Settings

To access general settings in the *Short-Circuit Analysis*, press the "*Short-Circuit Calculation*" command button located on the main menu. The important general settings to run *Short-Circuit Analysis* correctly are as follows:

I) **"Basic Options" tab**

Simulation method: DIgSILENT PowerFactory supports eight different methods to perform short circuit analysis:

1. *VDE 0102*: This method is defined by the German VDE standard. In this method, all system elements are considered in nominal terms. Therefore, this method is not suitable for large power networks with industrial loads.
2. *IEC 60909*: This method is defined by the International IEC standard. This method does not consider the installed capacitors in the power grid.
3. *ANSI*: This method is defined by the American ANSI/IEEE C37 standard.
4. *Complete*: In terms of analyzing different conditions, it is an ideal method. Therefore, all the performed analysis in this section will be based on this method.
5. *IEC 61363*
6. *IEC 61660 (DC)*: This method is defined by the International IEC standard and utilized for DC short circuit calculation.
7. *ANSI/IEEE 946 (DC)*: This method is defined by the ANSI/IEEE standard and utilized for DC short circuit calculation.
8. *DIN EN 61660 (DC)*

Available fault types:

• Three-phase short circuit • Two-phase short circuit • Single phase to ground • Two phase to ground • One phase to neutral • One-phase neutral to ground	• Two phase to neutral • Two-phase neutral to ground • Three phase to neutral • Three-phase neutral to ground • Three-phase short circuit (unbalanced)
• Multiple Faults: It can only be used if the "**complete**" option is selected.	

How to calculate: Short circuit calculations can be performed based on "Maximum Short-Circuit Currents" or "Minimum Short-Circuit Currents." Usually, the calculations are done based on the maximum short-circuit current mode. While the minimum short-circuit current mode is used for overcurrent relays.

Protection issues: Users can also consider the impact of fault impedance and short-circuit duration in computation concerning the conditions of the network under study.

Fault location:

In the **single** fault mode, the fault location can be defined in three different ways:

1. At User Selection: In this mode, for a single terminal, busbar, and line, *Short-Circuit Analysis* is carried out. An easier way to use the "User Selection" option is as follows:

Right-click on the desired element → "Calculate" section → "Short-Circuit" option

2. At Busbars and Junctions: For every terminal (*ElmTerm*) in the system, a short-circuit calculation is performed, independently.

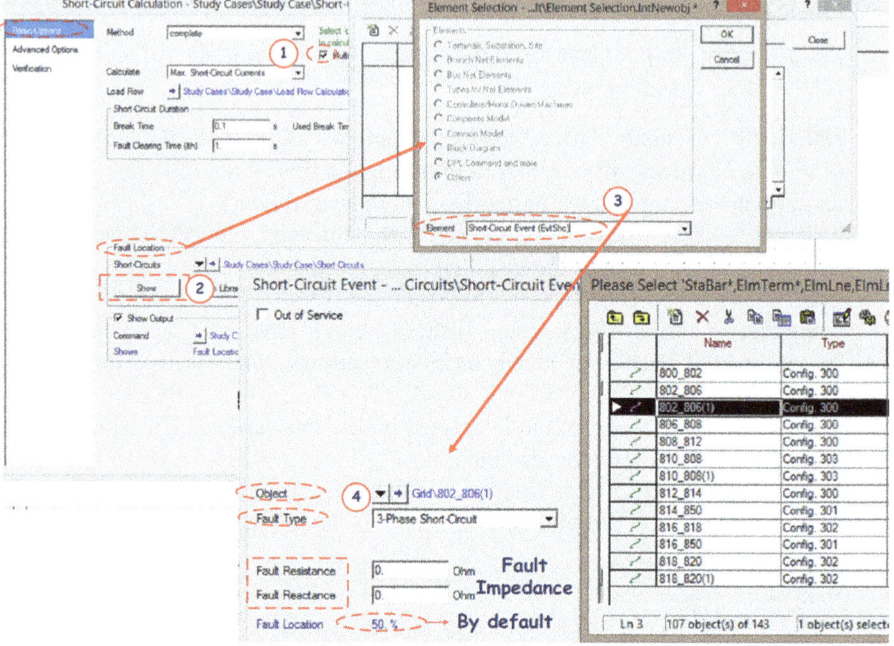

Fig. 5.1 Process of creating multiple fault mode

3. All Busbars: For each terminal (E*lmTerm*) configured as a busbar, a short-circuit calculation is performed, independently.

In the **multiple** fault mode, users can define multiple types of faults that occur simultaneously. To this end:

- *Enable the option "Multiple Faults"* (marked ❶ in Fig. 5.1);
- *Select the "Show" option from "Fault Location" box* (marked ❷ in Fig. 5.1);
- *Select the "New Object" option, and then choose "Short-Circuit Event (EvtShc)" using the "Element" drop-down list* (marked ❸ in Fig. 5.1);
- *Finally, choose the fault type and desired elements from the "Object" section* (marked ❹ in Fig. 5.1).

Another way to use this option is as follows:

Right-click on the desired element → "Calculate" section → "Multiple Faults" option → Close

It is important to note that by default, the fault location is set in the middle of each line. We can change the fault location as follows:

- *Double-clicking on the desired line*;
- *Go to the "Complete Short-Circuit" tab;*
- *Determine the desired location.*

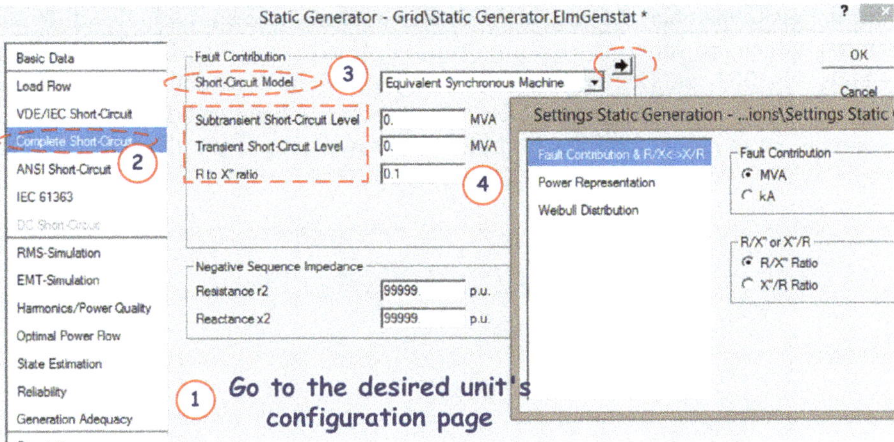

Fig. 5.2 Process of determining the short-circuit model for RES

5.1.1.2 Local Settings

The following methods can be used for short-circuit studies of RES in DIgSILENT PowerFactory:

I) **Equivalent synchronous machine**: To use this option, the user must have the following information about renewable sources:

- Subtransient short-circuit level.
- Transient short-circuit level.
- R to X'' ratio.
- Negative sequence impedance.

II) **Dynamic voltage support**: To use this option, the user must have the following information:

- Subtransient short-circuit level.
- R to X'' ratio.
- Maximum current (pu).

According to the conducted researches, most studies use the equivalent circuit method. The steps to access these methods are shown in Fig. 5.2.

5.1.2 Impact of RES on the Short-Circuit Power Level

In this part, the impact of RES on the short-circuit power level and the SCR index is evaluated via the *Short-Circuit Analysis* for different case studies. The 34-bus distribution network is chosen to perform the *Short-Circuit Analysis* in the presence of RES. The modified single-line diagram shown in Fig. 5.3. Using load and line data that are given in Tables 5.1, 5.2, and 5.3. The system's lines are set in five different modes, which are defined by Config. No. 300, Config. No. 301, Config. No. 302, Config. No. 303, and Config. No. 304. In addition, all lines are defined as the tower type (*TypTow*) mode. Resistance (R), reactance (X), and susceptance (B) matrices are required to model the lines in the form of tower type in the DIgSILENT. The

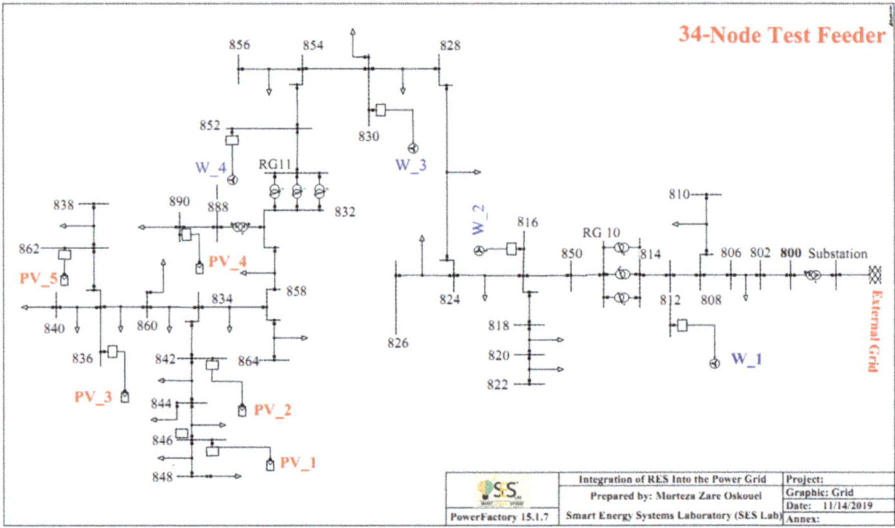

Fig. 5.3 Single-line diagram for IEEE 34-buses network

Table 5.1 Line segment data

Node A	Node B	Length (ft.)	Config.	Node A	Node B	Length (ft.)	Config.
800	802	2580	300	834	860	2020	301
802	806	1730	300	834	842	280	301
806	808	32,230	300	836	840	860	301
808	810	5804	303	836	862	280	301
808	812	37,500	300	842	844	1350	301
812	814	29,730	300	844	846	3640	301
814	850	10	301	846	848	530	301
816	818	1710	302	850	816	310	301
816	824	10,210	301	852	832	10	301
818	820	48,150	302	854	856	23,330	303
820	822	13,740	302	854	852	36,830	301
824	826	3030	303	858	864	1620	303
824	828	840	301	858	834	5830	301
828	830	20,440	301	860	836	2680	301
830	854	520	301	862	838	4860	304
832	858	4900	301	888	890	10,560	300

Table 5.2 Resistance, reactance, and susceptance matrices for tower type mode

Config.	*R* (ohms per mile)			*X* (ohms per mile)			*B* (micro Siemens per mile)		
300	0.2532	0.0398	0.0403	0.2527	0.1095	0.0950	1.0104	−0.2900	−0.1883
	0.0398	0.2507	0.0391	0.1095	0.2570	0.0870	−0.2900	0.9655	−0.1177
	0.0403	0.0391	0.2518	0.0950	0.0870	0.2551	−0.1883	−0.1177	0.9258
301	0.3655	0.0441	0.0447	0.2673	0.1220	0.1078	0.9698	−0.2720	−0.1781
	0.0441	0.3628	0.0433	0.1220	0.2705	0.0992	−0.2720	0.9291	−0.1127
	0.0447	0.0433	0.3640	0.1078	0.0992	0.2691	−0.1781	−0.1127	0.8931
302	0.5302	–	–	0.2813	–	–	0.8002	–	–
303	0.5302	–	–	0.2813	–	–	0.8002	–	–
304	0.3640	–	–	0.2692	–	–	0.8265	–	–

Table 5.3 Active and reactive power for distributed and spot loads

Node A	Node B	Ph-1 (kW)	Ph-1 (kvar)	Ph-2 (kW)	Ph-2 (kvar)	Ph-3 (kW)	Ph-3 (kvar)
802	806	0.00	0.00	89.25	44.63	74.38	41.65
808	810	0.00	0.00	47.60	23.80	0.00	0.00
818	820	101.15	50.58	0.00	0.00	0.00	0.00
820	822	401.63	208.25	0.00	0.00	0.00	0.00
816	824	0.00	0.00	14.88	5.95	0.00	0.00
824	826	0.00	0.00	119.00	59.50	0.00	0.00
824	828	0.00	0.00	0.00	0.00	11.90	5.95
828	830	20.83	8.93	0.00	0.00	0.00	0.00
854	856	0.00	0.00	11.90	5.95	0.00	0.00
832	858	20.83	8.93	5.95	2.98	17.85	8.93
858	864	5.95	2.98	0.00	0.00	0.00	0.00
858	834	11.90	5.95	44.63	23.80	38.68	20.83
834	860	47.60	23.80	59.50	29.75	327.25	163.63
860	836	89.25	44.63	29.75	17.85	124.95	65.45
836	840	53.55	26.78	65.45	32.73	0.00	0.00
862	838	0.00	0.00	83.30	41.65	0.00	0.00
842	844	26.78	14.88	0.00	0.00	0.00	0.00
844	846	0.00	0.00	74.38	35.70	59.50	32.73
846	848	0.00	0.00	68.43	32.73	0.00	0.00
860 (Spot Load)		59.50	47.60	59.50	47.60	59.50	47.60
840 (Spot Load)		26.78	20.83	26.78	20.83	26.78	20.83
844 (Spot Load)		401.63	312.38	401.63	312.38	401.63	312.38
848 (Spot Load)		59.50	47.60	59.50	47.60	59.50	47.60
890 (Spot Load)		446.25	223.13	446.25	223.13	446.25	223.13
830 (Spot Load)		29.75	14.88	29.75	14.88	74.38	29.75

Table 5.4 Wind and photovoltaic systems penetration details

	P (kW)	*Q* (kvar)	Subtransient short-circuit level (kA)	Transient short-circuit level (kA)	*R*/*X*″
Wind 1	450	85	10,000	9000	0.1
Wind 2	350	80	10,000	9000	0.1
Wind 3	450	95	10,000	9000	0.1
Wind 4	500	100	10,000	9000	0.1
PV 1	250	0	9000	7500	0.15
PV 2	300	0	9000	7500	0.15
PV 3	300	0	9000	7500	0.15
PV 4	220	0	9000	7500	0.15
PV 5	300	0	9000	7500	0.15

Table 5.5 Short-circuit power level in different busbars without considering RES

	Sk (MVA)		Sk_A (MVA)-Phase A	
Bus no.	3PH	3PH/0.5 Ω	1PH/0.1 Ω	1PH/0.5 Ω
Substation	3183.07	2194.74	3070.17	2194.8
800	5513.04	424.37	1905.04	428.37
808	22.16	21.13	17.57	17.11
812	**10.96**	**10.67**	**9.56**	**9.39**
816	**7.25**	**7.11**	**7.04**	**6.94**
822	2.74	2.72	2.73	2.72
828	6.35	6.24	6.5	6.42
830	**5.16**	**5.08**	**5.93**	**5.89**
836	**3.2**	**3.17**	**4.34**	**4.29**
842	**3.32**	**3.28**	**4.63**	**4.58**
846	**3.19**	**3.15**	**4.3**	**4.26**
852	**3.83**	**3.92**	**5.98**	**5.88**
860	3.27	3.24	4.51	4.46
862	**3.19**	**3.16**	**4.32**	**4.27**
864	4.89	4.82	4.87	4.82
890	**0.81**	**0.69**	**0.55**	**0.5**

used *R*, *X*, and *B* matrices are given in Table 5.2. The test system loads are modeled as unbalanced three phase (*PH-E*/*PH-D*) in the form of distributed loads and spot loads. The active and reactive power data of each phase are presented in Table 5.3. The capacity of each wind turbine and each photovoltaic (PV) system along with the required data to complete the "Short-Circuit" tab is listed in Table 5.4.

At first, the *Short-Circuit Analysis* is performed on the test system, regardless of the RES in four different statuses. It is advisable to run the short-circuit calculation under different conditions to evaluate the power grid status in different conditions. Table 5.5 presents the results of performing *Short-Circuit Analysis* based on the three-phase fault and single-phase fault (phase A) at all busbars. The short-circuit power level for the three-phase fault is determined under various ground resistance, which is set as 0 and 0.5 Ω. In addition, for the single-phase fault, the ground

Table 5.6 Short-circuit power level in different busbars with RES

	Sk (MVA)		Sk_A (MVA)-Phase A	
Bus no.	3PH	3PH/0.5 Ω	1PH/0.1 Ω	1PH/0.5 Ω
Substation	3189.14	2194.91	3071.57	2194.83
800	5523.91	424.28	1904.72	428.33
808	41.43	38.19	20.41	19.83
812	**143,782.8**	**391.14**	**14.17**	**13.86**
816	**143,783.6**	**388.57**	**11.49**	**11.28**
822	3.52	3.49	3.52	3.49
828	73.84	62.66	10.59	10.4
830	**143,775.68**	**367.05**	**11.15**	**10.93**
836	**39.02**	**35.9**	**9.09**	**8.94**
842	**54.4**	**48.43**	**10.03**	**9.85**
846	**36.62**	**33.79**	**8.95**	**8.81**
852	**143,778.76**	**339.31**	**14.51**	**14.09**
860	47.01	42.5	9.64	9.47
862	**38.28**	**35.27**	**9.04**	**8.89**
864	10.93	10.66	10.88	10.66
890	**3.99**	**3.06**	**1.04**	**0.96**

Table 5.7 Short-circuit ratio on busbars connected to RES

		SCR			
Bus no.	Pn (MW)	3PH	3PH/0.5 Ω	1PH/0.1 Ω	1PH/0.5 Ω
812 (W1)	0.45	319,517.3	869.2	31.5	30.8
816 (W2)	0.35	410,810.3	1110.2	32.8	32.2
830 (W3)	0.45	319,501.5	815.7	24.8	24.3
836 (PV3)	0.3	130.1	119.7	30.3	29.8
842 (PV2)	0.3	181.3	161.4	33.4	32.8
846 (PV1)	0.25	146.5	135.2	35.8	35.2
852 (W4)	0.5	287,557.5	678.6	29.0	28.2
862 (PV5)	0.3	127.6	117.6	30.1	29.6
890 (PV4)	0.22	18.1	13.9	4.7	4.4

resistance is set as 0.1 and 0.5 Ω. As can be seen in this table, the short-circuit power level decreases with increasing the ground resistance.

In the next step, short circuit studies are carried out in the presence of RES for the test system. The short-circuit power level for each busbar is calculated and shown in Table 5.6. As shown in Table 5.6, the short-circuit power level does not change at the high-voltage side (69 kV), but this parameter at the low-voltage side (24.9 kV) has increased dramatically in the presence of RES. Finally, the strength level of the test system at the RES interconnection point is calculated using Eq. (5.1). The SCR index for the desired busbars is presented in Table 5.7. According to the calculated SCR index, it is shown that the test system is a weak network and the misuse of RES can have devastating effects on the network under abnormal conditions.

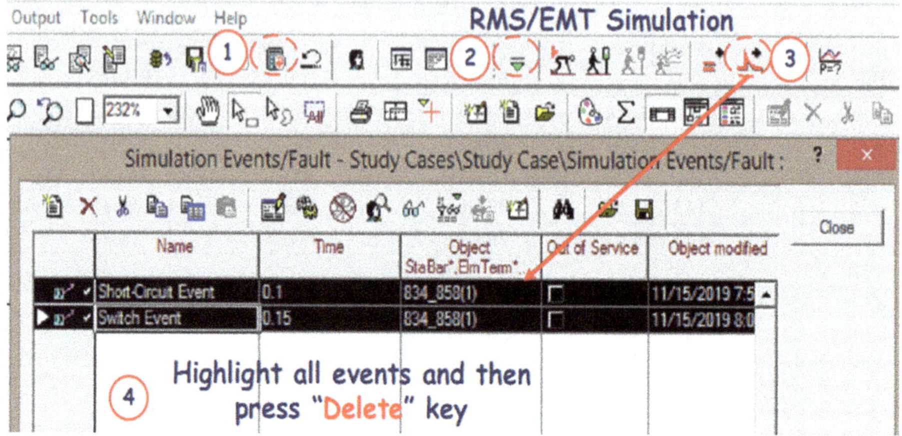

Fig. 5.4 Process of removing pre-defined events

5.2 Islanding Mode of Operation

The islanding mode of operation is an appropriate solution to maintain the vital components of power networks, such as the scheduled voltage and frequency within the allowable limit after an unforeseen event [6, 7]. In this mode, RES can help decision-makers for managing the power network resilience in extreme conditions. However, the use of RES may have a negative impact on the essential components of the network in the islanding mode [8, 9]. In this regard, *RMS Simulation* is the best analysis tool to evaluate the performance of the power network during islanding mode under various penetration levels of RES. The basic requirements for utilizing the *RMS Simulation* tool to investigate islanding mode in the power grid are as follows:

Before starting the RMS analysis:
At the first step: All the previous simulations must be cleared by the "Reset Calculation" command, which is located in the main toolbox (marked ❶ in Fig. 5.4).

At the next step: Selecting the "*RMS/EMT Simulation*" toolbar and then all previously defined events must be cleared through the "Edit Simulation Events" menu (marked ❸, ❹ in Fig. 5.4).

Step 1. Defining the new events: Users can define a new event in two ways:

1. Simple way: *Right-click on the desired element → Go to the "Define" option → Select one of the "Switch Event," "Short-Circuit Event," or "Planned Outage" options*

2. Professional way:

- *Select the "RMS/EMT Simulation" toolbar;*
- *Go to the "Edit Simulation Events" menu;*
- *In the opened window, select the "New Objective" button* (marked ❶ in Fig. 5.5);
- *Choose the considered events from the drop-down list* (marked ❷ in Fig. 5.5);

Available events:

• Dispatch Event (*EvtGen*)	• Short-Circuit Event (*EvtShc*)
• Load Event (*EvtLod*)	• Switch Event (*EvtSwitch*)
• Outage Event (*EvtOutage*)	• Synchronous Machine Event (*EvtSym*)
• Tap Event (*EvtTap*)	• External Measurement Event (*EvtExtmea*)
• Stop Event (*EvtStop*)	• Intercircuit Fault Event (*EvtShcll*)

- *Determine the target element for the selected event and enter the needed settings* (marked ❸, ❹ in Fig. 5.5).

Step 2. Defining the required objects for monitoring:

As shown in Fig. 5.6, an easier way to monitor each object of the power grid after performing *RMS Simulation* is as follows:

- *Right-click on the desired objects*;
- *Select "Results for RMS/EMT Simulation" option from the "Define" section* (marked ❶ in Fig. 5.6);

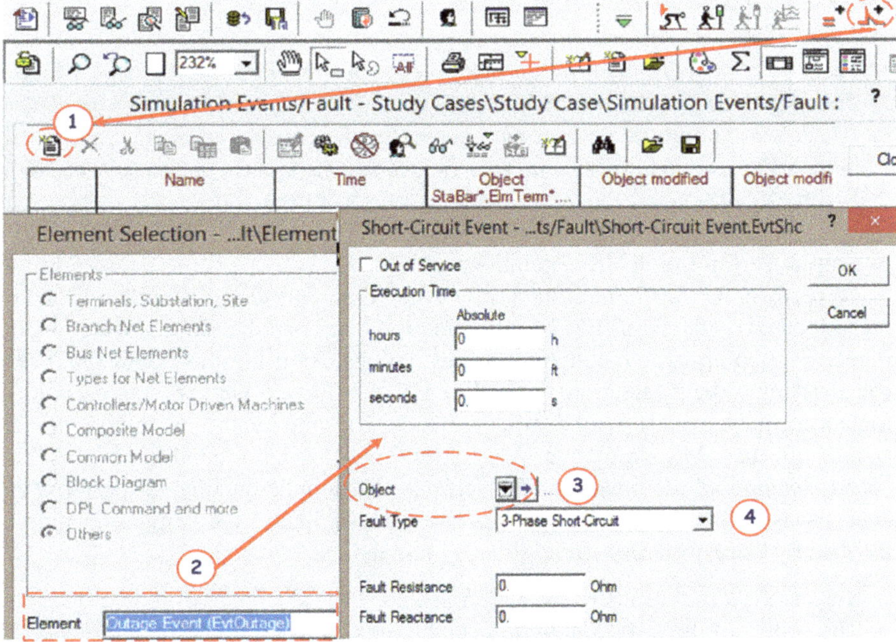

Fig. 5.5 Process of creating new events

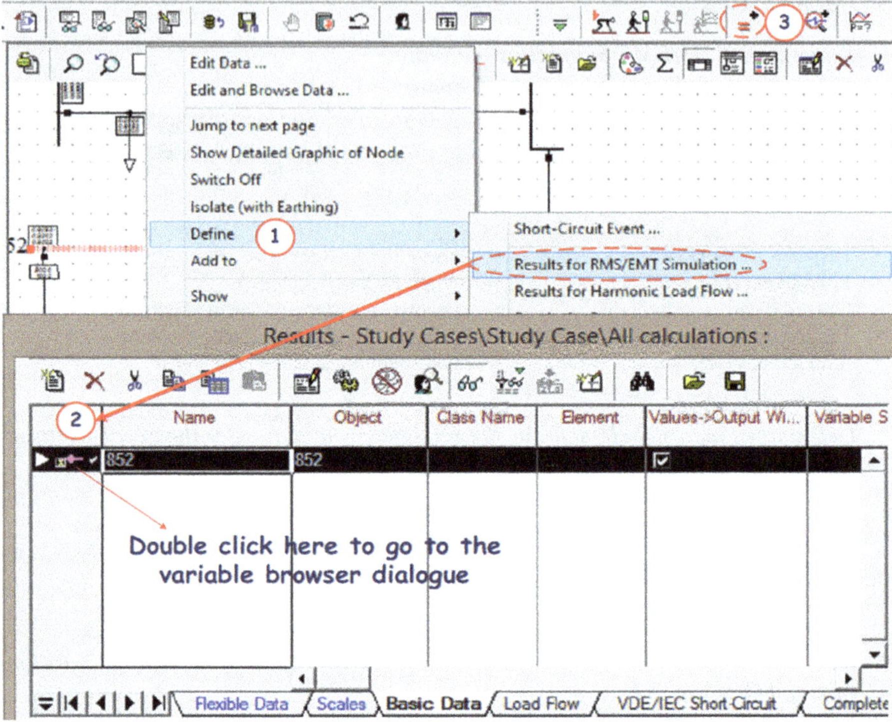

Fig. 5.6 Process of defining the variables for monitoring RMS simulation

- *Then, go to the variable browser dialogue to specify the considered variables for the selected object* (marked ❷ in Fig. 5.6).
- After determining the desired parameters for the considered objects, users can modify the selected items using the "*Edit Result Variables*" button (marked ❸ in Fig. 5.6).

Step 3. Performing RMS simulation:

To perform the *RMS Simulation*, press the "*Calculate Initial Conditions*" command located on the RMS/EMT Simulation toolbar. Then the settings window dialogue will pop-up, which contains various options. By using this command, users can determine the simulation method (RMS, EMT), the network representation (balanced, unbalanced), load flow commands, and maximum and minimum step size to run the RMS simulation. After specifying the initial conditions, press the "*Start Simulation*" command and determine the simulation time to perform the analysis. Note that users can interrupt the analysis before the simulation time is finished by using the "*Stop Simulation*" command.

Step 4. View the results of the analysis:

The results of the *RMS Simulation* can be seen by using the "*Create Simulation Plot*" icon, which is located on the *RMS/EMT Simulation* toolbar.

5.2.1 Assessment of Islanding Mode for 34-Bus System

The 34-bus distribution network as an appropriate test system is used to investigate the performance of the power grid with the presence of RES in the islanding mode. The islanding mode of operation can be evaluated in different modes depending on the strategic areas or critical areas of power grids. In this part, the performance of the network under study is evaluated when the three-phase faults occur at lines 852–854 and 834–842. After each fault occurs at the system's lines, the protection breakers are activated and the faulty line is isolated from the network. Accordingly, the following events are defined in the 34-bus distribution network to study the islanding mode:

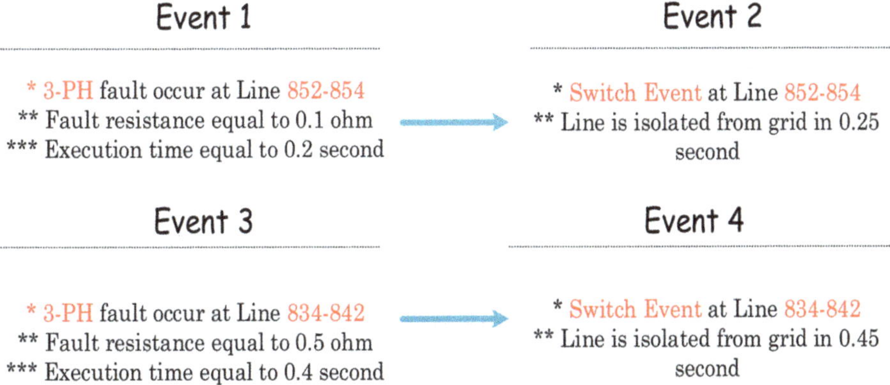

To evaluate the impact of RES on the system's performance and vital components, the *RMS Simulation* is implemented for three different cases. In **case 1**, defined events (Event 1–Event 4) for the test system are evaluated with 100% renewable energy system, whereas; in **case 2**, the total installed capacity of renewable energy is reduced to 75% of the nominal value and then the events are examined, eventually; in **case 3**, in addition to reducing the capacity of renewable sources, the network demands are also increased by 20% compared to the normal status, in which case the islanding mode of operation is evaluated in the peak period. The status of all network's buses should be examined in terms of voltage levels and frequency fluctuations according to each of the defined cases. For example, Figs. 5.7, 5.8, and 5.9 show the status of six different buses in terms of voltage level changes.

As shown in Fig. 5.7, after the first event occurred in 0.2 s, the second event occurred in 0.25 s (line 852–854 is isolated from the network), the voltage magnitude in the desired buses could increase by around 1.8 to 1.9 pu, while the maximum allowed voltage level is 1.05 pu. In addition, voltage began to oscillate with the occurrence of these events. Subsequently, after the third and fourth events occurred at line 834–842, the voltage level is decreased at the downstream buses but at the upstream buses increased again.

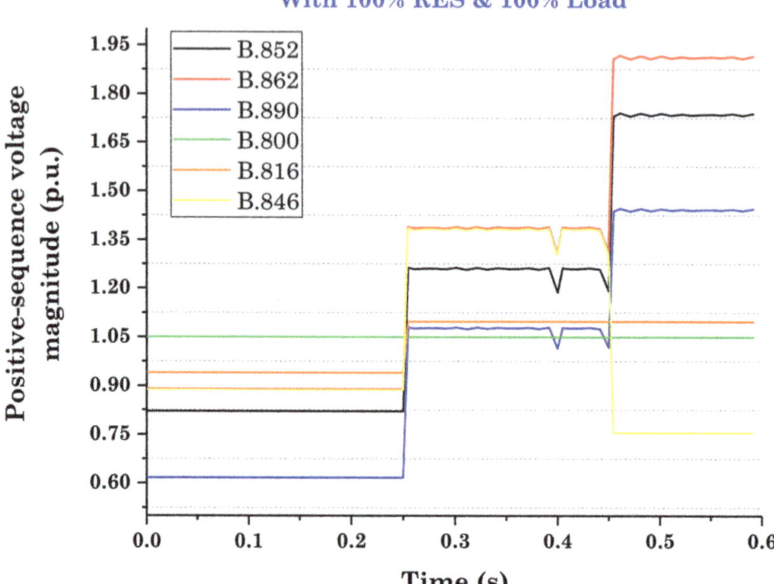

Fig. 5.7 Voltage variations during islanding mode for case 1

Active filters can be used to solve the problem of voltage waveforms. In general, although the use of RES during islanding mode brings some benefits, but RES can cause severe problems for the system operator in terms of power quality and unacceptable voltage level issues.

In the second and third case studies, the impacts of the installed capacity of RES, as well as the increase in demand, are investigated during islanding mode. According to Figs. 5.8 and 5.9, the behavior of the system after the events 1–4 occurred at lines 852–854 and 834–842 is exactly similar to case 1. Except that the voltage magnitude is reduced compared to the first case study at different buses. With the increasing demand rate, the voltage fluctuations also increased in various areas of the test system compared to the initial state.

5.3 Contingency Analysis

Contingency Analysis allows users to evaluate the impact of different events on the used elements in the power network. Some special contingencies may cause violations in network operating states, as well as cause instability conditions in the various zones of the power system. In other words, *Contingency Analysis* can be used to evaluate the security degree of power networks. Therefore, the performance of the power network must be analyzed by the system operator under both normal and

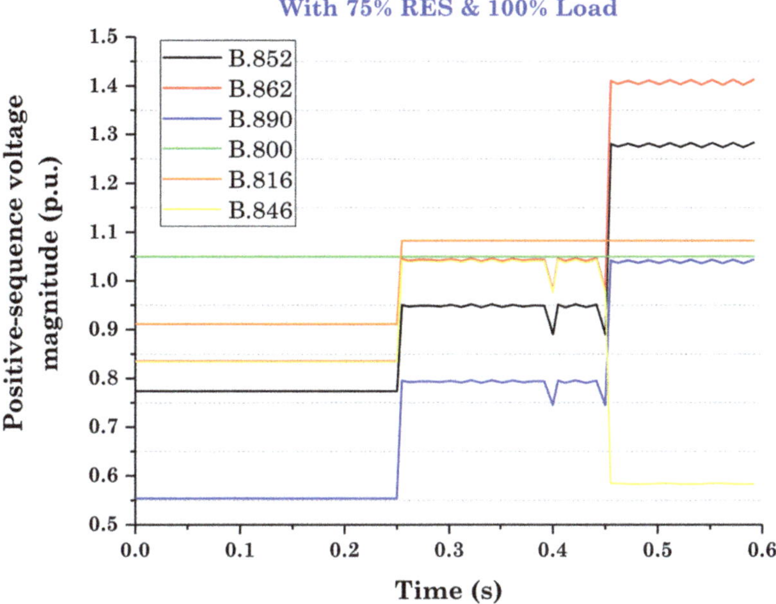

Fig. 5.8 Voltage variations during islanding mode for case 2

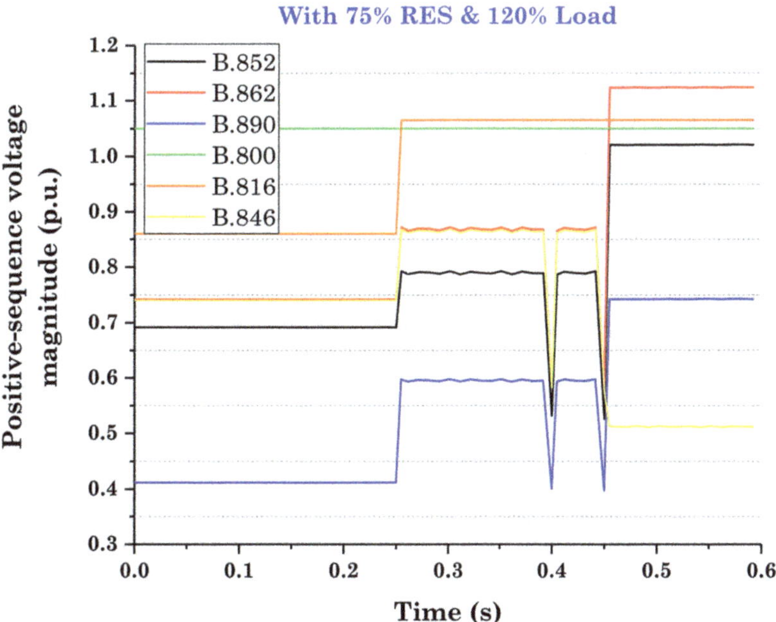

Fig. 5.9 Voltage variations during islanding mode for case 3

abnormal conditions with regards to the different fault cases. The system operator can provide alternative solutions to maintain the reliability constant at the standard level in daily operation by using this analysis. To this end, the maximum loading of the transmission lines and transformers, as well as the voltage level at each busbar, must be evaluated under the abnormal condition using pre-fault load and post-fault load flows.

Generally, *Contingency Analysis* consists of three steps:

Step 1. Defining contingency cases:

The various contingencies can be automatically generated using the "*Contingency Definition*" command available in the *Contingency Analysis* toolbar. Users can determine the desired settings for creating fault cases through this dialogue. Available options on this dialogue include:

- Creation of contingencies: The generated fault/contingency cases can be stored in the "*Operational Library*" (by using "*Generate Fault Cases for Library*" option) or can only be used for active study case without storing them in the library (by using "*Generate Contingency Cases for Analysis*" option).
- Outage level: Contingency analysis can be performed when an outage occurs in only one element ($n - 1$), or multiple outages occur in elements ($n - k$). Among the available options, ($n - 1$) outage creates single fault/contingency cases for each of the selected elements. The outage ($n - 2$) generates fault/contingency cases for every unique combination of two selected elements. Finally, ($n - k$) cases are available for every set of mutually coupled lines/cables. It should be noted that ($n - 2$) and ($n - k$) cases are specially used for strategic power networks.
- Network components: Contingency analysis can be done for both the whole system and user-defined selection of components. In the "Whole system" mode, fault/contingency cases can be created for all network elements (generators, transmission lines, transformers, and reactors) or predefined sets of elements with regards to the user's goals. In the "Selection" mode, fault/contingency cases can be created for a set of elements that are located in the user-defined zone. Figure 5.10 shows how to determine a user selection mode in contingency analysis.

By default, all events are defined as a three-phase short-circuit and the fault resistance is set to zero. From the "*Show Fault Cases/Groups*" command available in the *Contingency Analysis* toolbar, all defined fault/contingency cases are accessible (marked ❶, ❷ in Fig. 5.11). Users can modify the event type related to each element and define new events in the *Contingency Analysis* through the steps marked ❸–❻ in Fig. 5.11.

Step 2. Running the contingency analysis:

The *Contingency Analysis* performs the pre-fault and post-fault load flows for the defined fault/contingency cases and displays the required results based on the available variables. The "*Contingency Analysis*" command located on the simulation menu can be applied to perform this analysis. Then the command dialogue will

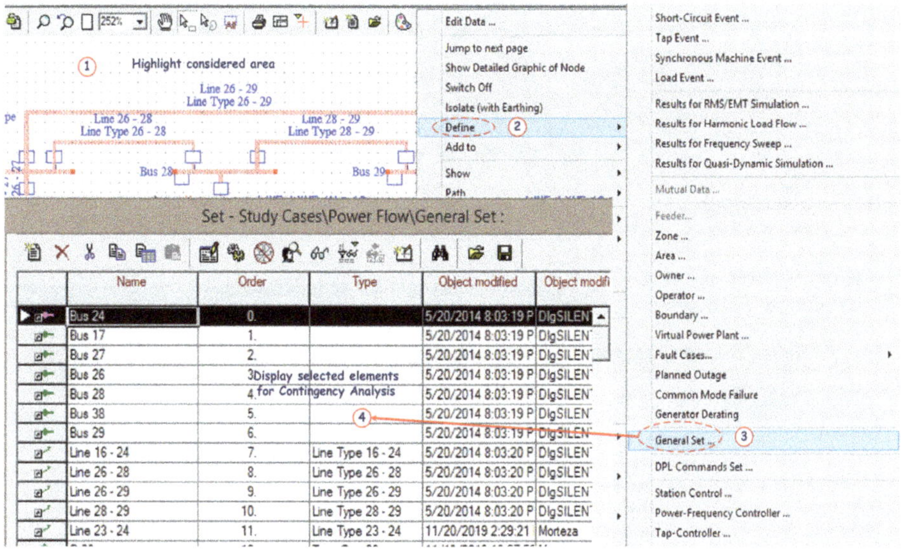

Fig. 5.10 Process of determining a user selection mode in contingency analysis

pop-up, which contains various configurations. The required settings to run the *Contingency Analysis* correctly are as follows:

I. **"Basic Options" tab**

Calculation method: The *Contingency Analysis* uses the AC or DC load flow methods to calculate the power flow based on the contingency cases. In large networks, DC load flow is often used to increase the computing speed. However, users can apply DC and AC load flow modes simultaneously for critical cases. In this state, at first, DC load flow mode is applied to calculate the active power flow according to the contingency cases. Then, the contingency simulation will recalculate the post-fault load flow using the AC load flow for certain contingencies that not within permitted boundaries.

Limits for recording: This section is used as a global filter to isolate the results of the *Contingency Analysis*. If one of the defined constraints is violated, the calculated result for each network component is recorded in the resulting dialogue. In PowerFactory, it is possible to define constraints for the thermal loading of components and allowed voltage level for busbars.

Contingencies: The determined fault/contingencycases to perform the simulation can be controlled through this section. From the "*Show*" button, all defined contingencies are accessible. The "*Add Cases/Groups*" button is used to manage

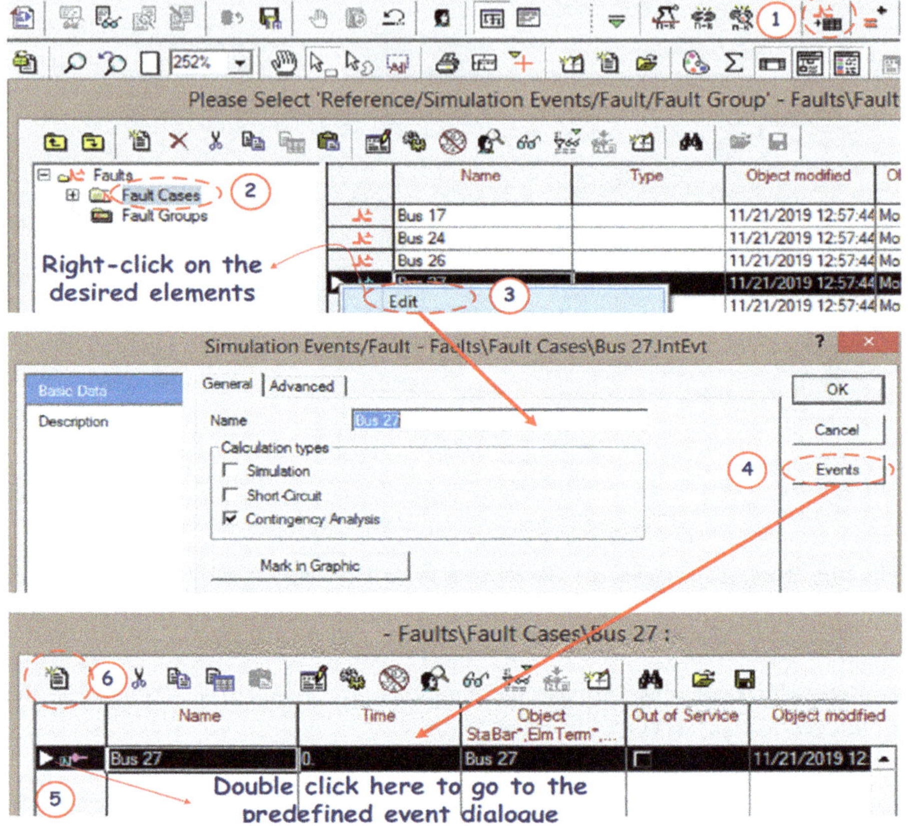

Fig. 5.11 Process of editing the defined fault cases in contingency analysis

defined contingency cases in the first step. All contingency cases stored in the *Contingency Analysis* command can be removed via the "*Remove All*" button.

II) **"Effectiveness" tab**

This tab allows the user to evaluate the effectiveness of each of the generating units (renewable sources or conventional generating units) in the post-fault mode of operation. It should be noted that the effectiveness mode is only available for the single time phase calculation.

III) **"Multiple Time Phases" tab**

Contingency Analysis in the form of DIgSILENT PowerFactory can be done in two different methods; (1) single time phase and (2) multiple time phase. In both methods, the pre-fault and post-fault load flow per contingency cases are compared to the specified loading and voltage constraints.

IV) **"Time Sweep" tab**

This option is used when the time characteristics are assigned to the system's loads or generating units. By using this tab, contingency analysis is automatically

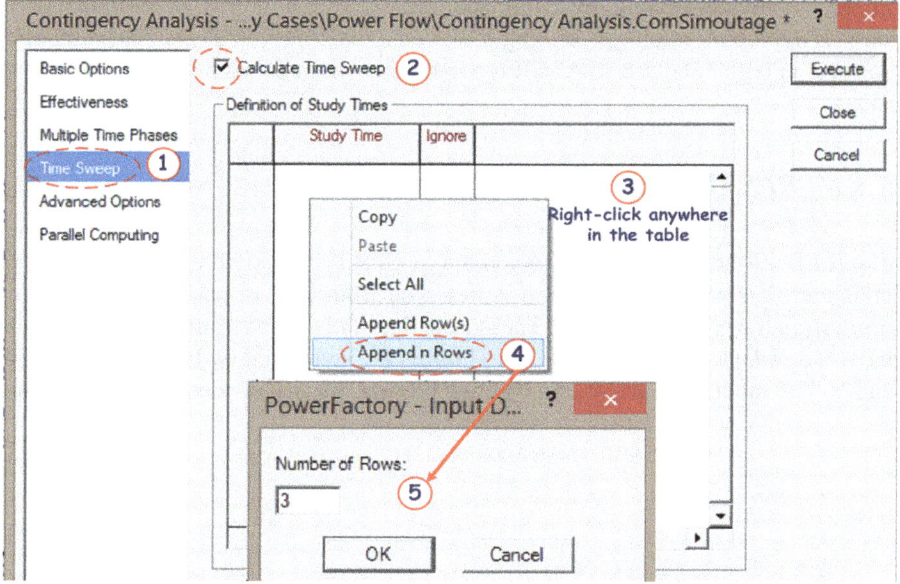

Fig. 5.12 Process of defining time sweep option in contingency analysis

performed for the predefined study times by the user. As shown in Fig. 5.12, the steps for assigning time sweep in the *Contingency Analysis* are as follows:

- *Enable the option "Calculate Time Sweep"* (marked ❷ in Fig. 5.12);
- *Right-click anywhere in the table then select "Append n Rows." After that allocate the number of needed rows* (marked ❸–❺ in Fig. 5.12);
- *Users can modify the date and time by double-clicking on the corresponding study time cell;*
- *Users can ignore the defined study times by enabling the "Ignore" flag.*

It should be noted that this tab is only available for the single time phase calculation.

After making all the desired settings and performing *Contingency Analysis,* PowerFactory may display an error message regarding the divergence issue in the *Load Flow Analysis*. However, this error does not cause any problems to run the *Contingency Analysis* correctly.

Step 3. Obtaining required outputs:

According to the defined fault/contingency cases, the results of the *Contingency Analysis* can be observed to investigate the performance of the power grid. Users can access the predefined outputs by clicking on the "*Report Contingency Analysis*

Results" command. Using this command, the software will provide various reports such as maximum loadings, loading violations, maximum voltages, and minimum voltages in both Tabular and ASCII reports based on the applied filters.

5.3.1 RES Operation in Contingency State

The IEEE 39-bus test system depicted in Fig. 5.13 is composed of ten conventional units and four wind farms, which is modified from [10] to perform *Contingency Analysis* in the presence of RES. The system data and the generating unit parameters have been adapted from [10, 11]. Wind farms are connected by busbars 38, 35, 32, and 9. The maximum active and reactive power capacity of wind turbine#1, wind

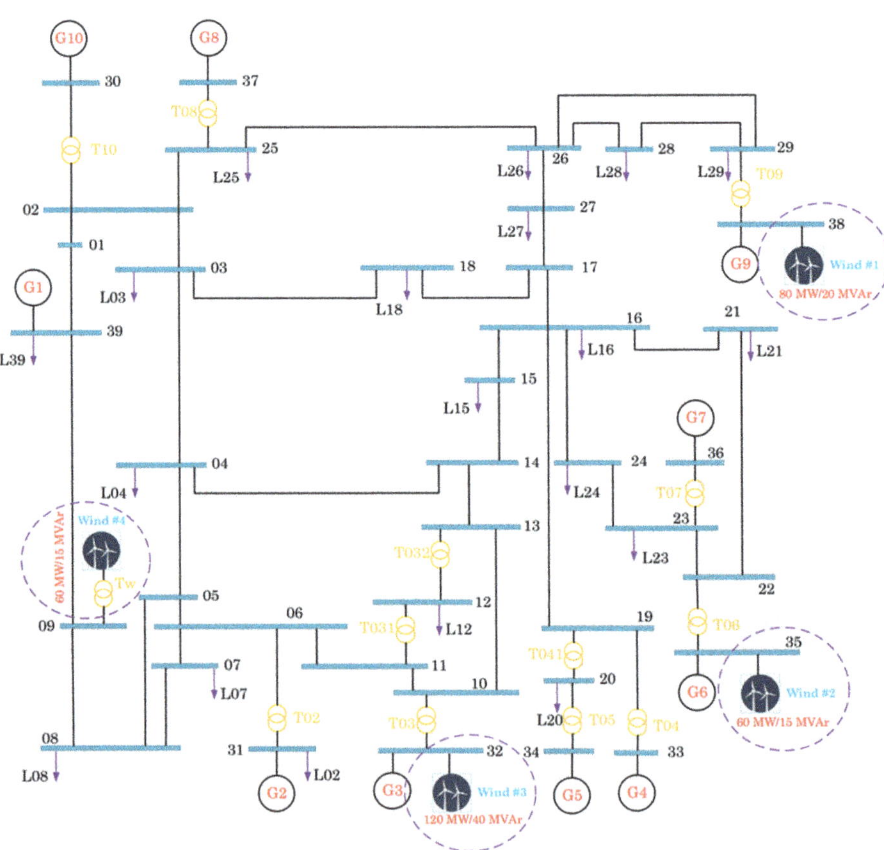

Fig. 5.13 Schematic diagram of the modified IEEE 39-bus system

Table 5.8 Contingency analysis report without considering wind farms: maximum loadings

Component	Loading Continuous [%]	Loading Base Case [%]	Contingency Name	Base Case and Continuous Loading
T02	211.37	79.58	T09	
Line 23 - 24	158.28	56.68	Line 21 - 22	
Line 21 - 22	156.92	99.08	Line 23 - 24	
Line 05 - 06	156.23	76.22	T09	
Line 06 - 07	123.46	71.88	Line 05 - 06	
Line 05 - 08	122.64	54.18	Line 06 - 07	
Line 16 - 21	111.80	53.76	Line 23 - 24	
Line 10 - 11	108.49	61.43	Line 10 - 13	
Line 22 - 23	108.40	12.04	Line 21 - 22	
T05	108.10	87.98	T041	
Line 10 - 13	107.97	47.45	Line 10 - 11	
Line 06 - 11	106.51	60.67	Line 13 - 14	
Line 13 - 14	106.09	45.69	Line 06 - 11	
Line 16 - 24	105.48	17.37	Line 21 - 22	
Line 16 - 19	101.39	81.90	T041	
Line 04 - 05	96.87	22.88	G 09	
Line 02 - 03	96.74	61.11	Line 26 - 27	
Line 04 - 14	95.30	45.75	Line 06 - 11	
Line 15 - 16	92.61	57.60	Line 16 - 17	
T03	91.13	86.67	G 09	

turbine#2, wind turbine#3, and wind turbine#4 are equal to (80 MW/20 MVAr), (60 MW/15 MVAr), (120 MW/40 MVAr), and (60 MW/15 MVAr), respectively.

To evaluate the role of RES in controlling abnormal conditions, the results of implementing *Contingency Analysis* on the IEEE 39-bus test system with and without considering wind farms are presented below. Hence, at first, the *Contingency Analysis* without the consideration of wind farms is performed. The numerical results for the maximum loading of the transmission lines and also the maximum voltage of the system's busbars considering the AC load flow are shown in Tables 5.8 and 5.9, respectively. These tables have five columns. The first column represents the components of the system. The next two columns give the maximum loadings and maximum voltages in the post-fault and pre-fault modes for the defined contingency cases, respectively. The results show that transformer T02 is overloaded (211.37%) when the transformer T09 is out of service. According to the network topology and simulation results, line 22–21, transformer T09, and conventional generating unit G09 play an important role in maintaining the test system in steady-state conditions.

Table 5.9 Contingency analysis report without considering wind farms: maximum voltages

Component	Max. voltage (pu)	Base voltage (pu)	Voltage step (pu)	Contingency name
Bus 24	1.08	1.04	0.04	Line 16–24
Bus 29	1.07	1.05	0.02	G 09
Bus 26	1.07	1.05	0.02	Line 26–27
Bus 28	1.06	1.04	0.02	G 09
Bus 01	1.06	1.05	0.02	Line 01–39
Bus 25	1.06	1.06	0.01	Line 26–27
Bus 36	1.06	1.06	0.00	Base case
Bus 19	1.06	1.05	0.01	Line 16–19
Bus 02	1.06	1.05	0.01	Line 02–03
Bus 23	1.05	1.04	0.01	Line 16–24
Bus 22	1.05	1.05	0.00	Line 15–16

In the next step, the *Contingency Analysis* is carried out considering the wind farms to investigate the variations of the maximum loading of the lines and the maximum voltage of the system's busbars under abnormal conditions. Numerical results in Tables 5.10 and 5.11 show the improvement of loading rates on various transmission lines and the voltage level at each busbar in the presence of the wind farms under contingency conditions. For example, under post-fault mode per contingency cases, the loading rate on transformer T02 has decreased from 211.37 to 177.91% when the transformer T09 is out of service. Due to the high penetration of wind power, the role of conventional generating units (G01–G10) in the contingency conditions has been reduced.

5.4 Evaluation of Wind Power Curtailment With and Without Considering BESS

As reported in the literature, one of the most efficient approaches to meet demand in power networks is renewable sources [12]. However, there is not enough infrastructure, such as line capacity, for using high penetration levels of RES in the power networks. For this reason, the renewable energy spillage rate in power grids has increased over the past years [13, 14]. In the absence of proper planning for using high-level RES, serious problems such as line congestion and voltage deviation will occur in the power networks. For more efficient use of RES in the power networks, innovative concepts such as high-efficient energy storage [15] and reconfigurable systems were presented [16]. Some studies have focused on finding appropriate solutions for the utilization of high-level RES using the battery energy storage system (BESS). The main targets of the presented solutions are to harvest higher wind energy, manage line congestion, and decrease voltage deviation [17–19].

Table 5.10 Contingency analysis report with wind farms: maximum loadings

Component	Loading Continuous [%]	Loading Base Case [%]	Contingency Name	Base Case and Continuous Loading [0 % - 178 %]
T02	177.91	37.27	T09	
Line 23 - 24	169.18	60.17	Line 21 - 22	
Line 21 - 22	167.12	105.41	Line 23 - 24	
Line 05 - 06	141.39	53.86	T09	
Line 16 - 21	121.90	60.09	Line 23 - 24	
Line 10 - 13	121.52	41.49	Line 10 - 11	
Line 10 - 11	121.25	80.49	Line 10 - 13	
Line 13 - 14	119.71	38.16	Line 06 - 11	
Line 06 - 11	119.33	81.21	Line 13 - 14	
Line 22 - 23	118.60	14.07	Line 21 - 22	
Line 04 - 14	117.48	50.17	Line 06 - 11	
Line 16 - 24	115.85	18.70	Line 21 - 22	
Line 05 - 08	111.10	49.91	Line 06 - 07	
Line 02 - 03	110.73	67.37	Line 26 - 27	
Line 06 - 07	110.68	64.28	Line 05 - 08	
T05	108.10	88.09	T041	
Line 26 - 29	107.70	41.51	Line 28 - 29	
Line 28 - 29	107.38	65.61	Line 26 - 29	
Line 15 - 16	101.88	68.04	Line 16 - 17	
Line 16 - 19	101.70	82.15	T041	
Line 26 - 27	100.86	53.47	Line 02 - 25	
Line 02 - 25	99.92	50.33	Line 26 - 27	
Line 16 - 17	97.91	35.41	Line 15 - 16	
T09	94.41	92.73	Line 28 - 29	
Line 17 - 18	91.45	41.64	Line 15 - 16	
T06	91.08	88.82	Line 23 - 24	
T03	90.48	86.37	Line 15 - 16	

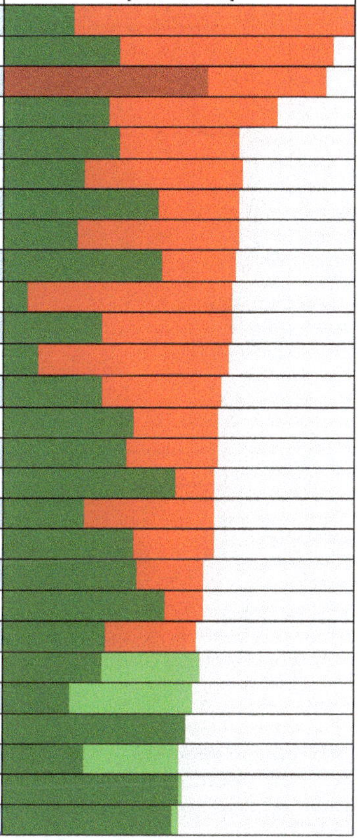

Table 5.11 Contingency analysis report with wind farms: maximum voltages

Component	Max. voltage (pu)	Voltage step (pu)	Base voltage (pu)	Contingency name
Bus 24	1.07	0.03	1.04	Line 16–24
Bus 26	1.07	0.02	1.05	Line 26–27
Bus 01	1.07	0.02	1.05	Line 01–39
Bus 25	1.06	0.01	1.06	Line 26–27
Bus 36	1.06	0.00	1.06	Base case
Bus 19	1.06	0.01	1.05	Line 16–19
Bus 28	1.06	0.01	1.05	Line 26–27
Bus 02	1.06	0.01	1.05	Line 02–03
Bus 29	1.06	0.01	1.05	Line 26–27
Bus 23	1.05	0.01	1.04	Line 16–24
Bus 22	1.05	0.00	1.05	Line 15–16

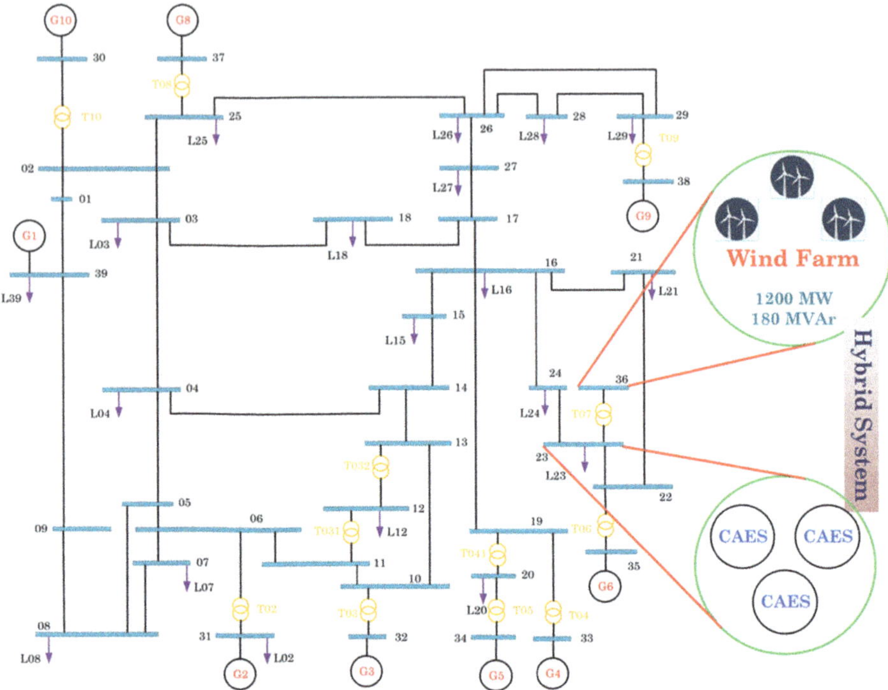

Fig. 5.14 Electrical diagram of the IEEE 39-bus system along with the hybrid system

Energy storage technologies have the potential to enable power networks to deal with the system anomalies, for example, reduction of the duration of power outages [20], minimization of active and reactive power losses, minimization of the voltage deviation, and enhancement of the power quality [21].

In this part, DIgSILENT PowerFactory is applied to investigate the phenomenon of wind power spillage in the power system with and without considering BESS. To this end, the IEEE 39-bus system is chosen as a case study, which is widely used in the studies for testing purposes. To evaluate the wind power curtailment rate, we use a wind farm instead of the generator G07 (at bus 36) in the IEEE 39-bus test system. The maximum power capacity of the wind farm is equal to 1200 MW. On the other hand, novel technologies such as compressed air energy storage (CAES) system can be used as the promising technologies for reducing power spillage of wind farms, compensate for the uncertainty in the output power of wind turbines, and increasing energy efficiency. Hence, three CAES systems, which are located at bus 23 were used alongside the wind farm. The test system is depicted in Fig. 5.14. This example demonstrates the usefulness of the DIgSILENT PowerFactory to develop power grids based on RES and BESS.

By existing standards, the maximum allowable loading rate on the transmission lines is 85%. Therefore, if the loading rate on each line exceeds this value, the system operator should improve the network conditions using various methods, including

Table 5.12 Technical comparison of network features when using different size of BESS

Wind power curtailment (MW)	CAES performance (MW)	Line loading (%)		Allowed to exploit?
		Line 23–24	Line 16–21	
0	0	107.38	110.09	NO
280	0	84.68	84.88	YES
0	−100 (*Charging mode*)	99.3	101.109	NO
200	−100 (*Charging mode*)	83.2	83.23	YES
0	−200 (*Charging mode*)	91.3	92.2	NO
100	−200 (*Charging mode*)	83.31	83.3	YES
0	**−350** (*Charging mode*)	**79.53**	**79.12**	**YES**

reducing the output power of generating units. In the face of renewable sources, this action is in direct conflict with the interests of the network operator. For the considered test system, the role of lines 16–21 and 23–24 in transferring generated power by the wind farm to other parts of the network is more than other lines. To evaluate the amount of wind power spillage, we first perform *Load Flow Analysis* on the test system without considering CAES systems. All technical results under different conditions are reported in Table 5.12. According to this table, it can be seen that loading rates on transmission lines 23–24 and 16–21 without using storage resources are equal to 107.38% and 110.09%, respectively, which are outside the permissible range (loading rates >85%). As a result, 280 MW of wind power must be curtailed to reduce the loading rate on these two lines. In the next step, different capacities of the storage system are examined to reduce the curtailment of wind power.

To prevent wind power curtailment, the hybrid system operator can store surplus power in the CAES system as compressed air. Then, the energy stored in the CAES system can be used to supply the power network's consumption load. The amount of wind power curtailment for different capacity levels of the CAES system according to the line loading level is given in Table 5.12. According to the numerical results given in Table 5.12, by using CAES systems with the nominal capacity of 350 MWh, the curtailment of wind power is equal to zero and lines 23–24 and 16–21 reach a standard level of loading. The positive powers of the CAES system correspond to the intervals of discharging action and the negative powers correspond to the intervals of charging operation. Besides, the use of storage systems can help the network operator to execute demand response programs (DRPs) and increase the spinning reserve in the power network.

References

1. Wu D, Li G, Javadi M et al (2018) Assessing impact of renewable energy integration on system strength using site-dependent short circuit ratio. IEEE Trans Sustain Energy 9(3):1072–1080
2. Huang SH, Schmall J, Conto J, et al (2012) Voltage control challenges on weak grids with high penetration of wind generation: ERCOT experience. Paper presented at the IEEE Power and Energy Society General Meeting, pp 1–7

3. Urdal H, Ierna R, Zhu J et al (2015) System strength considerations in a converter dominated power system. IET Renew Power Gener 9(1):10–17
4. Zhou Y, Nguyen DD, Kjaer PC, et al (2013) Connecting wind power plant with weak grid–challenges and solutions. Paper presented at the IEEE Power and Energy Society General Meeting, pp 1–5, July
5. Transmission and Distribution Committee of the IEEE Power Engineering Society (1997). IEEE guide for planning DC links terminating at AC locations having low short-circuit capacities. IEEE standard n.1204–1997
6. Xu X, Yan Z, Shahidehpour M et al (2019) Maximum loadability of islanded microgrids with renewable energy generation. IEEE Trans Smart Grid 10(5):4696–4705
7. Singh R, Bansal RC, Singh AR et al (2018) Multi-objective optimization of hybrid renewable energy system using reformed electric system cascade analysis for islanding and grid connected modes of operation. IEEE Access 6:47332–47354
8. Delgado-Antillón CP, Domínguez-Navarro JA (2018) Probabilistic siting and sizing of energy storage systems in distribution power systems based on the islanding feature. Electr Power Syst Res 155:225–235
9. Youssef KH (2017) Power quality constrained optimal management of unbalanced smart microgrids during scheduled multiple transitions between grid-connected and islanded modes. IEEE Trans Smart Grid 8(1):457–464
10. Zheng L, Hu W, Lu Q et al (2015) Optimal energy storage system allocation and operation for improving wind power penetration. IET Gener Transm Distrib 9:2672–2678
11. Zimmerman RD, Murillo-Sanchez CE, Thomas RJ (2011) MATPOWER: steady-state operations, planning, and analysis tools for power systems research and education. IEEE Trans Power Syst 26:12–19
12. Zare Oskouei M, Sadeghi Yazdankhah A (2015) Scenario-based stochastic optimal operation of wind, photovoltaic, pump-storage hybrid system in frequency-based pricing. Energy Convers Manag 105:1105–1114
13. Abbaspour M, Satkin M, Mohammadi-Ivatloo B et al (2013) Optimal operation scheduling of wind power integrated with compressed air energy storage (CAES). Renew Energy 51:53–59
14. Aliasghari P, Zamani-Gargari M, Mohammadi-Ivatloo B (2018) Look-ahead risk-constrained scheduling of wind power integrated system with compressed air energy storage (CAES) plant. Energy 160:668–677
15. Zare Oskouei M, Sadeghi Yazdankhah A (2017) The role of coordinated load shifting and frequency-based pricing strategies in maximizing hybrid system profit. Energy 135:370–381
16. Hemmati M, Ghasemzadeh S, Mohammadi-Ivatloo B (2018) Optimal scheduling of smart reconfigurable neighbour micro-grids. IET Gener Transm Distrib 13(3):380–389
17. Jadidbonab M, Mousavi-Sarabi H, Mohammadi-Ivatloo B (2018) Risk-constrained scheduling of solar-based three state compressed air energy storage with waste thermal recovery unit in the thermal energy market environment. IET Renew Power Gener 13(6):920–929
18. Ghahramani M, Nazari-Heris M, Zare K et al (2019) Energy and reserve management of a smart distribution system by incorporating responsive-loads/battery/wind turbines considering uncertain parameters. Energy 183:205–219
19. Karimi A, Aminifar F, Fereidunian A et al (2019) Energy storage allocation in wind integrated distribution networks: an MILP-based approach. Renew Energy 134:1042–1055
20. Arif A, Wang Z, Wang J et al (2018) Power distribution system outage management with co-optimization of repairs, reconfiguration, and DG dispatch. IEEE Trans Smart Grid 9(5):4109–4118
21. Bollen MHJ, Das R, Djokic S et al (2017) Power quality concerns in implementing smart distribution-grid applications. IEEE Trans Smart Grid 8(1):391–399

Chapter 6
Reliability Assessment in the Presence of Renewable Energy Sources

This chapter presents the instruction on how to evaluate the reliability indices in the presence of renewable energy sources (RES) in DIgSILENT PowerFactory. In this chapter, the reliability issue is examined from two perspectives. In the first step, the impacts of RES on the improvement of various reliability indices are investigated. In the next step, how to calculate the fundamental reliability indices in the renewable power generation sector, such as loss of load probability (LOLP), is presented. To this end, different case studies are analyzed and the various reliability indices are assessed on the standard test system. The conceptual architecture of the considered analyses in this chapter is presented in Fig. 6.1.

6.1 Power System Reliability Issues

The main task of the power grid is to meet customers' needs with adequate, stable, and reliable power. Reliability indices are the most appropriate tool for evaluating the performance of the power grid by the system's operator [1, 2]. Reliability assessment is used to measure expected interruption frequency, outage duration, annual interruptions costs, and system availability to compare different network designs with each other [3, 4].

In practical researches, the evaluation of reliability indices in the real power systems with consideration of various natural or technical events has attracted much attention from the researchers' perspective. Generally speaking, the power system reliability issue has a vast range of studies, and some of them are summarized as follows:

- A practical method to determine the reliability indices of radial and loop power distribution networks was presented in [5], considering the capabilities of

M. Zare Oskouei, B. Mohammadi-Ivatloo, *Integration of Renewable Energy Sources Into the Power Grid Through PowerFactory*, Power Systems,
https://doi.org/10.1007/978-3-030-44376-4_6

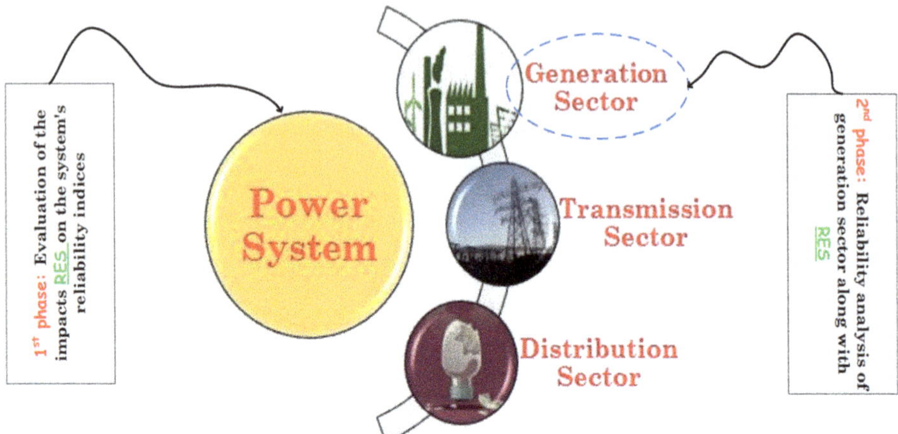

Fig. 6.1 Conceptual architecture

DIgSILENT PowerFactory. This method is capable of examining different indexes in terms of subscribers as well as the system operator.

- In [6, 7], the effects of the very small power producer and active power filters on power system reliability were investigated by PowerFactory. In these researches, the main concern is to improve the system's reliability as well as decrease the power system losses relying on the high penetration level of renewable energy sources (RES).
- In [8], a comprehensive model was developed to integrate photovoltaic systems and energy storage into an existing microgrid. The main focus of this work is to improve the system's reliability when various failures occur in the network.

According to the IEEE-P1366 standard [9], the basic indicators for system *Reliability Analysis* are evaluated from the perspective of network and subscribers based on momentary interruptions and sustained interruptions. These indicators help decision-makers to monitor technical and management measures to promote the system's performance. Important indices are presented in the following subsections for power system reliability analysis.

6.1.1 Important Reliability Indices

There are some necessary reliability indices to assess power system adequacy and security. The applications and mathematical formulations of each of these indices are presented below:

- **Whole system reliability calculations**

The most commonly used indices to assess the reliability of the power grid are system average interruption frequency index (SAIFI), system average interruption duration index (SAIDI), customer average interruption duration index (CAIDI), customer average interruption frequency index (CAIFI), average service availability index (ASAI), and energy not supplied (ENS). These indices can be calculated as follows [10–12]:

The SAIDI index represents the average time that customers' power is interrupted during the time period in 1 year.

$$\mathrm{SAIDI} = \frac{\sum_i (N_i \times F_i)}{N_T}; \qquad \mathrm{Unit}: h / C \cdot a \tag{6.1}$$

where N_i is the number of affected customers interrupted at faults, N_T is the total number of serving customers, F_i is the restoration time after occurred each fault.

The CAIDI index represents the average time to restore service and its mathematical relation is similar to the SAIDI index.

$$\mathrm{CAIDI} = \frac{\sum_i (N_i \times F_i)}{\sum_i N_i}; \qquad \mathrm{Unit}: h \tag{6.2}$$

The SAIFI index demonstrates how often interruption occurs on average for each customer during the time period under study. This index is defined as follows:

$$\mathrm{SAIFI} = \frac{\mathrm{SAIDI}}{\mathrm{CAIDI}}; \qquad \mathrm{Unit}: 1 / C \cdot a \tag{6.3}$$

The CAIFI index represents the average number of sustained interruptions per customer interrupted per year. Only subscribers who are experiencing sustained interruptions are involved in the calculation of this metric. This index is calculated as follows:

$$\mathrm{CAIFI} = \Sigma \frac{N_o}{N_i}; \qquad \mathrm{Unit}: 1 / C \cdot a \tag{6.4}$$

where N_o is the number of interruptions.

The ASAI index represents the ratio of the total number of customer hours that service was available during the defined calculation period to the total customer hours demanded. This index is calculated as follows:

$$\mathrm{ASAI} = \left[1 - \frac{\sum_i (N_i \times F_i)}{N_T \times T} \right] \times 100 \tag{6.5}$$

where T is the time period under study.

The ENS represents the average amount of energy not delivered to the customers. This index is calculated as follows:

$$\text{ENS} = \sum_i \left(L_i \times U_i\right); \qquad \text{Unit}: MWh/a \tag{6.6}$$

where L_i is the average load demand and U_i is the annual outage time of the load point i.

- **Generation system reliability calculations**

The loss of load probability (LOLP) is an important index to evaluate generation system reliability. This index is based on the probabilistic approaches and refers to the time when the generating capacity is not able to cover the system demand. The LOLP can be calculated according to the annual load duration curve or the daily load duration curve. The mathematical formulation for the calculation of LOLP is given in Eq. (6.7).

$$\text{LOLP} = \frac{\sum_j \left(P_j \times t_j\right)}{100} \tag{6.7}$$

where P_j is the probability of occurrence of the peak load equal to or greater than the generation capacity at time t_j.

- **Load point reliability calculations**

In this context, load point interruption frequency (LPIF) and load point interruption time (LPIT) indicators are the most important metrics for evaluating system performance in meeting customer needs. These indices are calculated as follows:

$$\text{LPIF}_i = \sum_k \text{FL}_k \qquad Unit: 1/a \tag{6.8a}$$

$$\text{LPIT}_i = \sum_k 8760 \cdot \rho_k \qquad \text{Unit}: h/a \tag{6.8b}$$

where FL_k is the frequency of occurrence of contingency k and ρ_k is the probability of occurrence of contingency k at load point i.

6.2 Fundamental Settings to Evaluate System Reliability

Reliability issues can be assessed from both network adequacy and security point of view. It is necessary to model different events that cause network disruptions to specify the level of network reliability. Transmission line outages, the occurrence of various faults at the system's busbars, and different problems related to transformers

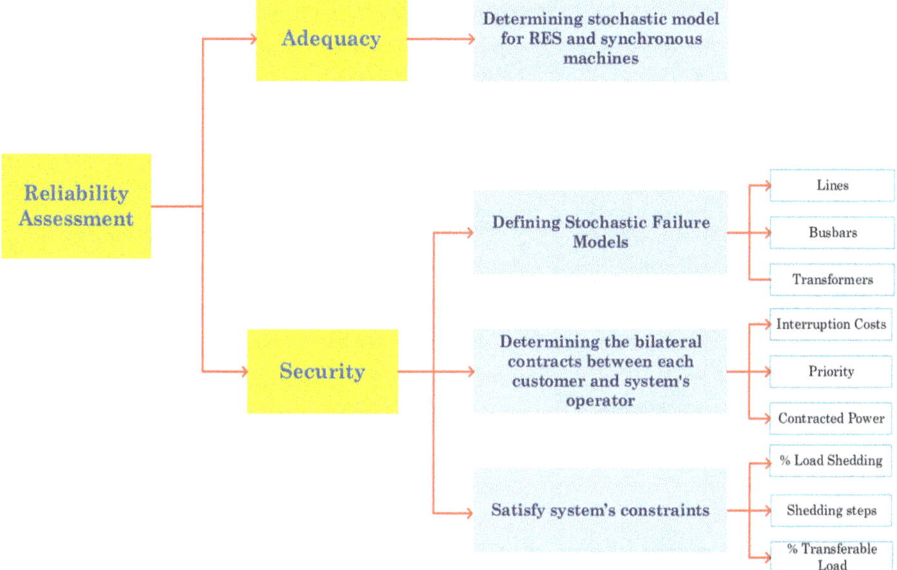

Fig. 6.2 Reliability assessment process

are essential phenomena that disrupt the power system. The reliability issue in the power system can be assessed by two important parameters, namely "mean time to failure (**MTTF**)" and "mean time to repair (**MTTR**)," which must be defined for all equipment. According to Eq. (6.9), these parameters are inversely related to failure and repair rates.

$$\mathrm{MTTF} = \frac{1}{\lambda} \tag{6.9a}$$

$$\mathrm{MTTR} = \frac{1}{\mu} \tag{6.9b}$$

where λ is the constant failure rate and μ is the constant repair rate.

The various steps to evaluate the reliability of the power system are illustrated in Fig. 6.2. The required settings to perform the *Reliability Analysis* are described below.

6.2.1 Stochastic Failure Model for Lines, Buses, and Transformers

The definition of a stochastic failure model for lines, buses, and transformers is similar and has the same steps. For this reason, the required settings for buses are provided below, which can be generalized to other equipment:

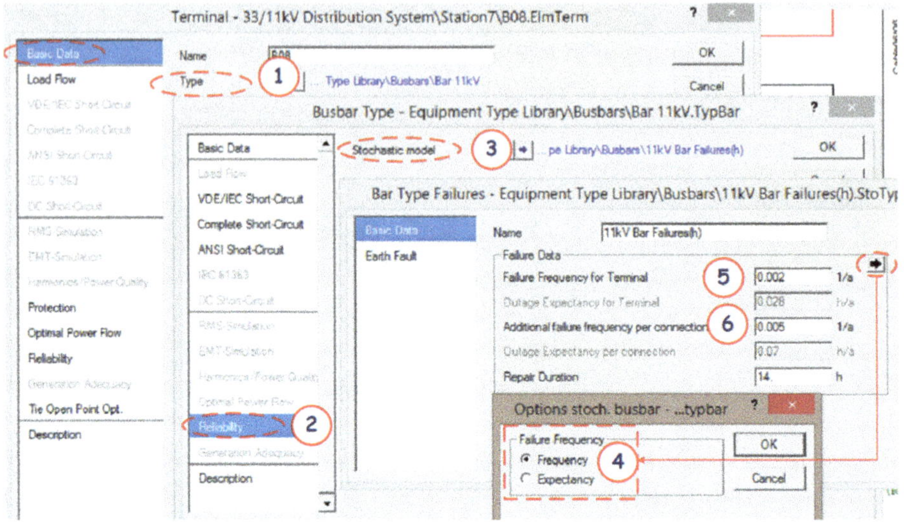

Fig. 6.3 Required settings to assess reliability indices

- *Double-clicking on the desired bus.*
- *Go to the "Basic Data" tab and select "New Project Type-Line Type" from "Type" section* (marked ❶ in Fig. 6.3);
- *Go to the "Reliability" tab after determining appropriate type for the busbar* (marked ❷ in Fig. 6.3);
- *Select "New Project Type" from "Stochastic Model" section* (marked ❸ in Fig. 6.3);
- *In the opened window, go to the "Basic Data";*
- *Users can be determined stochastic model based on "Frequency" or "Expectancy"* (marked ❹ in Fig. 6.3);
- *In the frequency mode, the user must be a determined failure (λ) and repair (μ) rates according to the historical information for each equipment* (marked ❺ in Fig. 6.3).

The relationship between parameters included in the "Failure Data" box is as follows:

Repair Duration × Failure Frequency for Terminal = Outage Expectancy for Terminal

Each busbar has several lines or transformers. Users can be applied the "Additional failure frequency per connection" option to consider the effect of the number of connections on each busbar (marked ❻ in Fig. 6.3). Generally, the following relationship is established for each busbar:

Forced Outage Rate = Failure Frequency for Terminal + (No. of connection × Additional failure frequency per connection)

In addition to the failure data, reliability indices can be obtained based on the short circuit capability. To this end:

- *Go to the "Earth Fault" tab* (marked ❶ in Fig. 6.4);
- *Enable the option "Model Earth Faults"* (marked ❷ in Fig. 6.4);
- *Determine the data for the "Frequency of single earth faults," "Conditional probability of a second earth fault," and "Repair Duration" in hours* (marked ❸ in Fig. 6.4).

It should be noted that following the earth fault on a component may also occur the insulation breakdown on another component, which may lead to the second simultaneous earth fault. These steps are shown in Figs. 6.3 and 6.4.

- *After applying desired settings, Press OK to return to the element dialogue;*
- *Go to the "Reliability" tab to view the entered information;*
- *Disable the option "Ideal component."*

All the above mentioned steps must be repeated for all busbars. If there are several special busbars whose failure data differ from other busbars, the following steps can be taken to determine the stochastic failure model:

- *Double-clicking on the desired busbar;*
- *Go to the "Reliability" tab and disable the option "Ideal component";*
- *Select "New Project Type" from "Element Model" section;*
- *The rest of the steps are as described in the previous mode.*

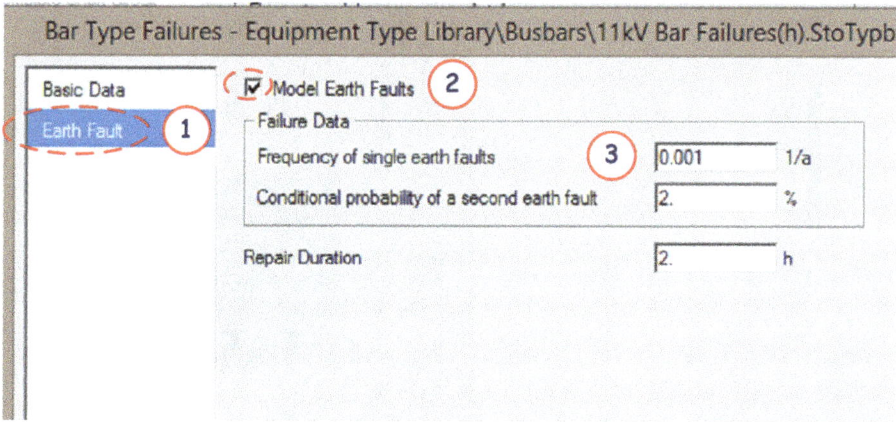

Fig. 6.4 Process of defining earth fault in reliability analysis

6.2.2 Modeling of General Loads for Reliability Calculation

In this section, the required configuration related to general loads is presented to increase the accuracy of *Reliability Analysis*. To this end:

- *Double-clicking on the desired loads and then go to the "Reliability" tab.*

In the reliability page, the **first item** is related to the number of customers connected to the desired load. As shown in Fig. 6.5, at different voltage levels, each load can represent several substations or several subscribers. Determining the exact number of subscribers is very effective in calculating various indices such as SAIFI and CAIFI.

The **second item** is related to the load reduction. The load reduction on the power system is intended to satisfy various network constraints such as voltage or thermal constraints. The overload phenomenon can be managed by either shedding or transferring some of the loads. According to the network under study, the appropriate method can be used to reduce the required demand. The features of the various networks and how to deal with the overload issue in these networks are shown in Fig. 6.6. Hence, to determine the status of each subscriber to deal with the overload phenomenon, the following setting can be done in the "Reliability" page for each general load:

- Determining the "load priority" using the "*Priority*" box: Operator tries to shed or transfer the loads with the lowest priority first. Therefore, it is advisable to allocate high values for strategic loads; for example 100.

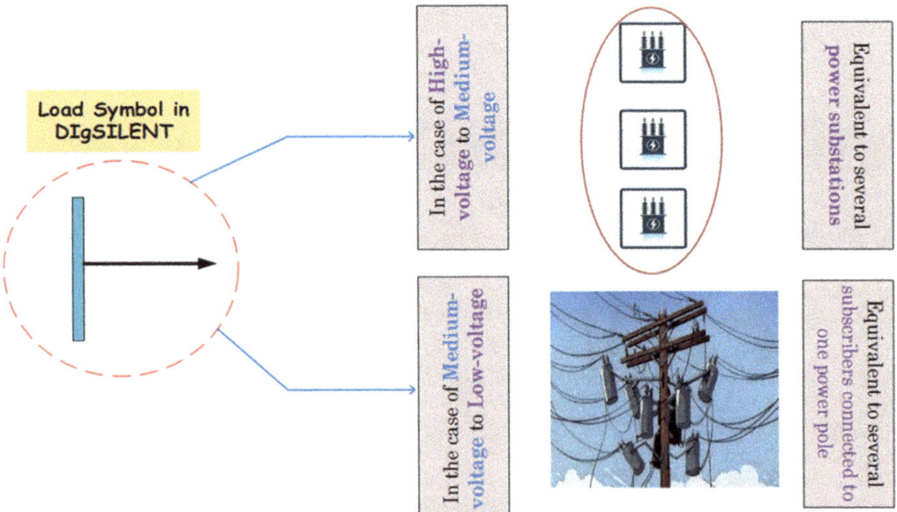

Fig. 6.5 Integrating several subscribers in DIgSILENT

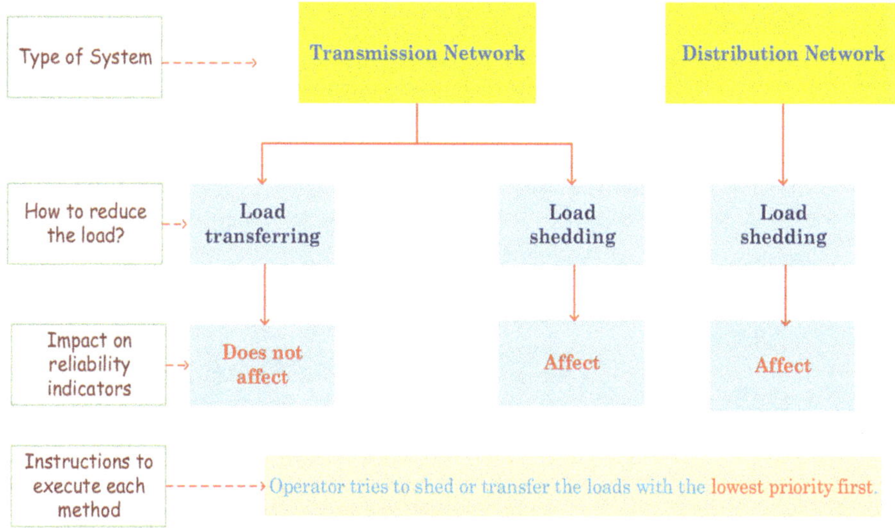

Fig. 6.6 Implementation of different load reduction methods in power system

- Determining the "number of load shedding steps" using the "*Shedding steps*" box: The ideal mode for the system operator is to use the infinite option. In this case, each load demand can be shed to the desired value to alleviate the system's constraints. In other shedding steps modes, for example, five shedding steps mean that the load can be shed to 20, 40, 60, 80, or 100% of its base value. In the restructured electricity market environment, the best state to shed the load is to use the largest number of steps. Note that the amount of load shed per customer will directly affect the reliability indicators.
- Determining the "load transfer percentage" using the "*Transferable*" box: The determined percentage for each subscriber can be transferred from the current network to the neighbor network. Note that, the transferred load is considered as a supplied load, whereas shed load is a unserved load. Therefore, the amount of load transferred does not affect the reliability indices. In addition, users can use an alternative source to supply the transferred load. The "*Alternative Supply (Load)*" option can be used to do this.

The **third item** is related to interruption costs paid to the customers to implement load shedding program. In DIgSILENT PowerFactory, this cost can be modeled by "Energy Tariff" and "Time Tariff" curves. To this end, various actions like Fig. 6.7 must be taken.

- *Go to the "Interruption Costs" box* (marked ❶ in Fig. 6.7);
- *Choose the "Select" option from the "Tariff" section* (marked ❷ in Fig. 6.7);
- *In the opened window, select the "New Objective" button in the "Parameter Characteristic" folder* (marked ❸ in Fig. 6.7);
- *The tariff type can be chosen from the pop-up window* (marked ❹ in Fig. 6.7).

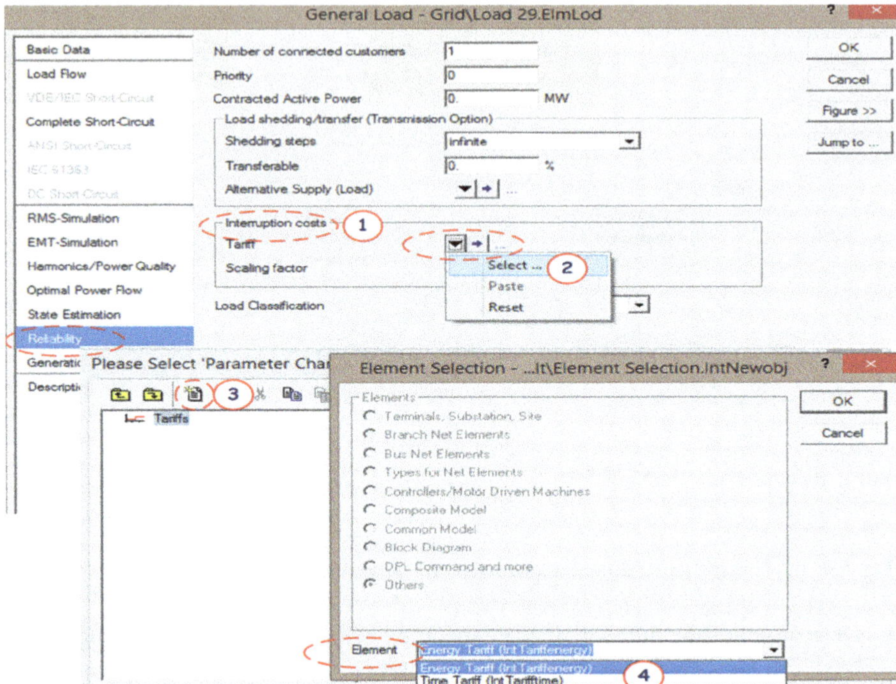

Fig. 6.7 Process of determining energy tariff and time tariff curves

6.2.3 *Modeling of Synchronous Machine for Reliability Calculation*

Synchronous generators can affect reliability indicators using re-dispatching actions. According to Fig. 6.8, the "Active power steps," "Priority," and "Active power operational limits" tools must be determined to specify the best generators to re-dispatch. Each unit can vary its production capacity to maintain constraints based on different steps (marked ❶ in Fig. 6.8). Also, the system operator tries to change the generation power of the lowest priority units first (marked ❷ in Fig. 6.8). These changes should be within the allowable range of each unit (marked ❸ in Fig. 6.8).

6.2.4 *Common Mode Failures*

A common mode failure represents the various failure of two or more power grid components at the same time. For example, when there is a problem with a transformer, then the generation unit connected to the transformer will be interrupted simultaneously. Therefore, these components have a common failure mode. To

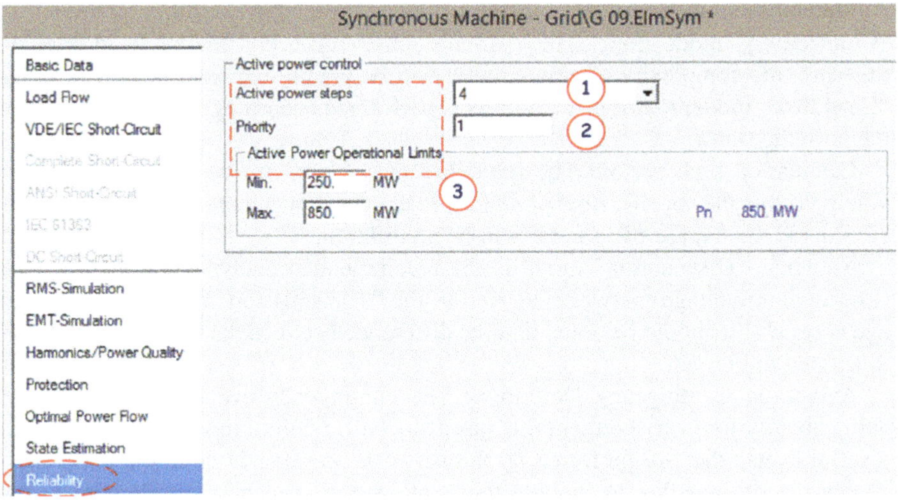

Fig. 6.8 Required configuration of synchronous machines in reliability analysis

define a common failure mode in PowerFactory, users must follow the steps given below:

- *Hold down* ***Ctrl*** *and select the desired components you want to create a common failure mode;*
- *Right-click on the selected components then choose "Common Mode Failure" from "Define" option;*
- *Press OK to save changes.*

The defined modes are saved in the following path, which can be edited or deleted from the "Common Mode Failures" folder:

Data manager → Active project's library → Operational library → Common mode failures

6.3 Implementation of Reliability Analysis

Reliability Analysis can be performed after applying the described items in the previous sections to all the components in the system under study. To perform the *Reliability Analysis*, press the "*Calculation*" command located on the main menu, and select *Reliability Analysis*. Then the settings dialogue will pop-up, which contains various options. The important settings to run reliability simulation correctly are as follows:

I. **"Basic Options" tab**

Calculation section: Users can determine the network's type (balanced or unbalanced) and the requirements for performing *Load Flow Analysis* from this section.

Method section: *Reliability Analysis* can be performed in two different modes. In "Connectivity" mode, none of the system's constraints is considered. In addition, in this case, the complexity of computation will be greatly reduced. However, in the "Load flow" mode, *Reliability Analysis* is performed according to all limitations of the power system.

Calculation time period: The reliability calculation can be performed for the whole year or the specific time of year. If load curves or maintenance plans are not considered, then these two modes will be no different.

Network: "Distribution" mode is the recommended analysis option for radial (low/medium voltage) networks as well as the "Transmission" mode for high voltage networks. It should be noted that the difference between the two options is fully illustrated in Fig. 6.6.

Automatic Contingency Definition: The contingency events can be selected automatically for every element that has a stochastic failure model or manually for a set of events that are preferred by the network operator. In addition, users must enable the desired flag to include the contingency state of each element in the *Reliability Analysis*.

II. **"Costs" tab**

Using this tab, users can assign energy tariffs to the amount of energy not supplied. In addition, a global or individual cost curve can be defined for all loads.

III. **"Constraints" tab**

The maximum allowable thermal loading of lines, the allowable range for terminals' voltage, and the voltage drop of feeders must be determined using this tab. It is best to apply all constraints using the global constraint for all components. Typically, the maximum thermal loading for all lines is equal to 85%, the lower voltage limit for all terminals is equal to 0.95 pu, and the maximum voltage drop for the selected feeders is equal to 5%.

IV. **"Maintenance" tab**

Using this tab, users can define planned outages to investigate the role of different equipment outages in the results of the *Reliability Analysis*. How to define and use this option is described in more detail in Chap. 3.

V. **"Load Data" tab**

This tab will be activated if the reliability calculation option "Complete Year" is chosen on the basic options tab. From this tab, "Load states" or "Load distribution states" options can be considered in the *Reliability Analysis*. To use this tab, the load state must first be created based on the defined time-based characteristics for the system's loads. To this end:

- *Click the "Create Load States" button from the "Reliability Analysis" toolbar* (marked ❶ in Fig. 6.9);
- *Determine the time period for the calculation of load states as well as the "Accuracy" of load state calculation. Users can create more load states by assigning a lower accuracy percentage.* (marked ❷ in Fig. 6.9);

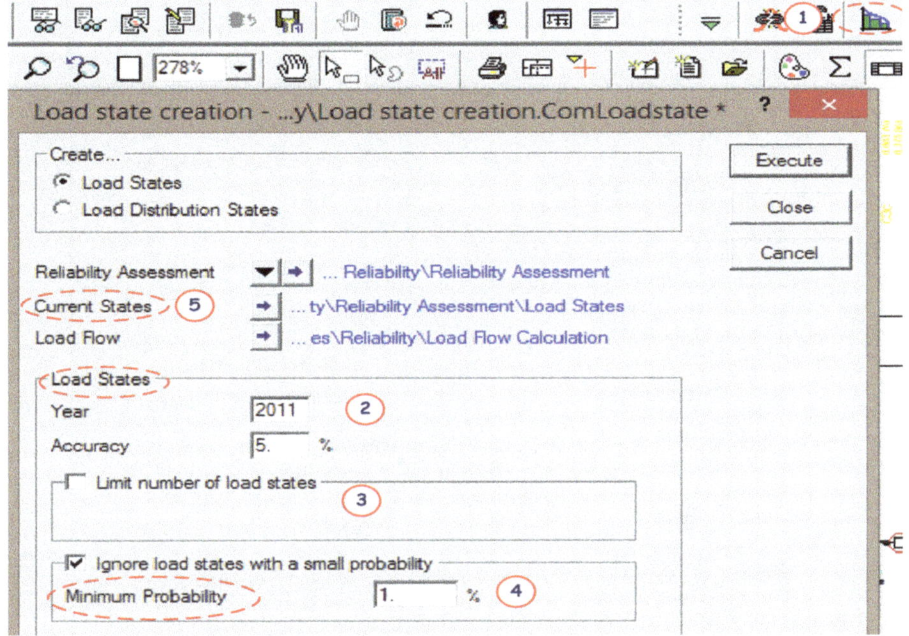

Fig. 6.9 Process of load states

- *Users can limit the number of created load states* (marked ❸ in Fig. 6.9);
- *In addition, users can define a threshold for ignoring load states with a low probability* (marked ❹ in Fig. 6.9).

View the created load state:

- *Go to the "Create Load States" menu from the "Reliability Analysis" toolbar*;
- *Go to the "Current States" section* (marked ❺ in Fig. 6.9);
- *In the opened window, you can see created clusters for each load and probability plots for each customer.*

6.3.1 Impact of RES on Reliability Indices

In this part, a sample 33/11 kV distribution test system is used to assess the effect of RES on the reliability indices. Figure 6.10 shows the test system topology. The fundamental data of the test system, including the length of lines and general load characteristics, are given in Tables 6.1 and 6.2. The test system is modeled in the balanced mode and the type of general loads is considered 3PH (PH-E). Table 6.3 presents the input data for the stochastic failure model for busbars, transformers, and lines. In addition, the interruption cost for the test system's loads in the form of time tariff is presented in Table 6.4. As shown in Fig. 6.11, the load scaling factor is

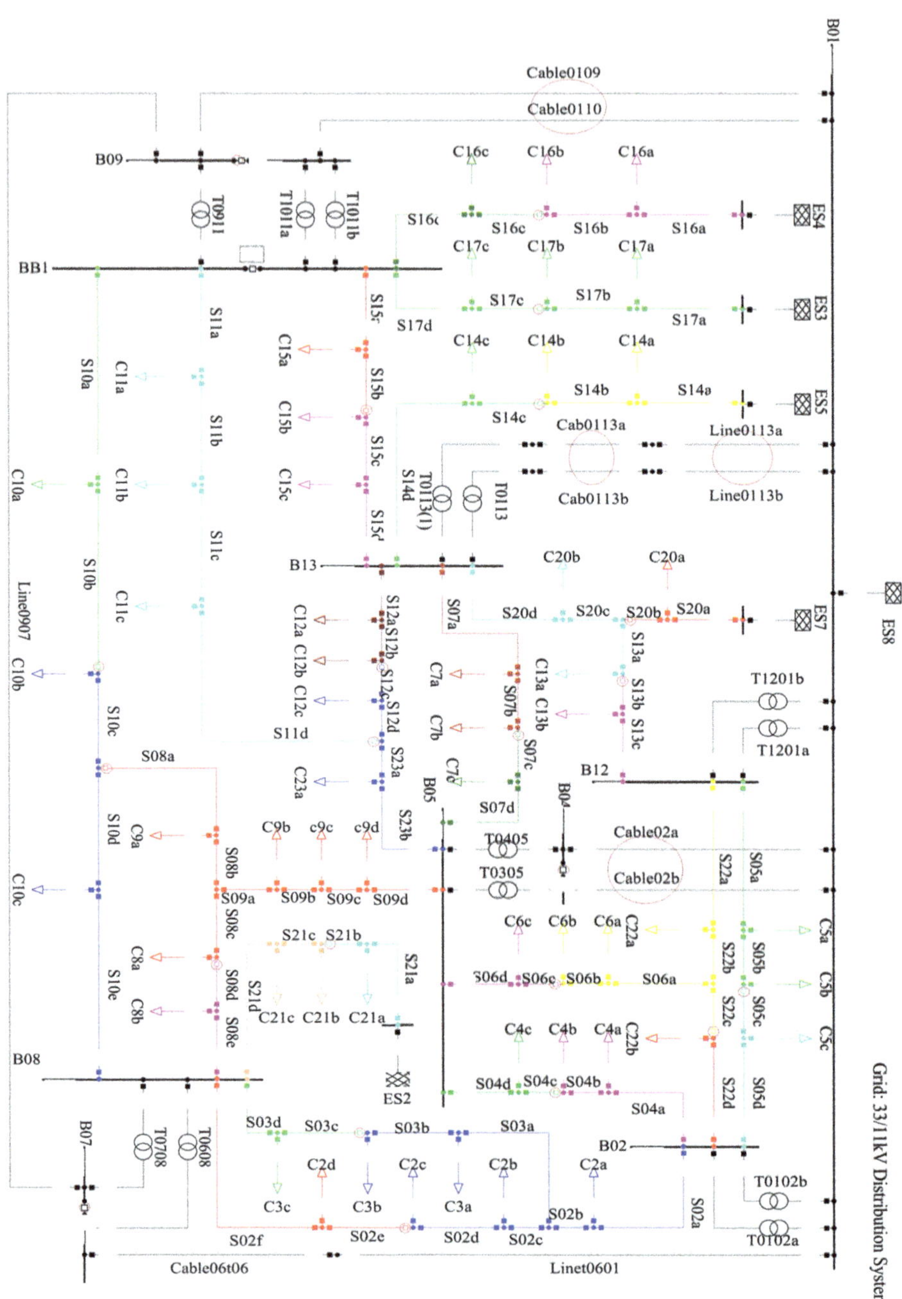

Fig. 6.10 Electrical diagram of the test system

Table 6.1 The length of grid lines

Name	Length (km)	Name	Length (km)	Name	Length (km)	Name	Length (km)
Cab0113a	6	S04c	1.5	S10a	6	S15d	1.5
Cab0113b	6	S04d	1.5	S10b	2.4	S16a	4.8
Cable0109	27.84	S05a	1.5	S10c	4.8	S16b	1.2
Cable0110	27.84	S05b	1.5	S10d	4.8	S16c	0.6
Cable02a	11.76	S05c	1.5	S10e	3	S16d	1.8
Cable02b	11.76	S05d	1.5	S11a	0.75	S17a	4.8
Cable06t06	6	S06a	1.5	S11b	0.75	S17b	1.2
Line0113a	11.76	S06b	1.5	S11c	0.75	S17c	0.6
Line0113b	11.76	S06c	1.5	S11d	0.75	S17d	1.8
Line0907	27.12	S06d	1.5	S12a	1.5	S20a	1.5
Linet0601	36.24	S07a	1.5	S12b	0.75	S20b	1.5
S02a	3	S07b	1.5	S12c	0.375	S20c	1.5
S02b	1.5	S07c	1.5	S12d	0.375	S20d	1.5
S02c	0.375	S07d	1.5	S13a	3	S21a	1.5
S02d	0.375	S08a	2.4	S13b	1.5	S21b	1.5
S02e	0.375	S08b	1.2	S13c	1.5	S21c	1.5
S02f	0.375	S08c	0.6	S14a	2.4	S21d	1.5
S03a	1.5	S08d	1.8	S14b	0.12	S22a	1.5
S03b	1.5	S08e	0.6	S14c	3.6	S22b	1.5
S03c	1.5	S09a	1.5	S14d	0.6	S22c	1.5
S03d	1.5	S09b	1.5	S15a	1.5	S22d	1.5
S04a	1.5	S09c	1.5	S15b	1.5	S23a	3
S04b	1.5	S09d	1.5	S15c	1.5	S23b	3

used to evaluate the reliability indices with regard to the hourly load changes. The cost allocated to energy not supplied is one of the most important parameters needed to calculate reliability indices. Hence, the cost function used in this analysis is shown in Fig. 6.12.

Two cases are considered to assess the reliability indices. The first case performs the *Reliability Analysis* without considering RES alternatives. The second case considers the role of RES in changing reliability indices.

Case I: Investigating the reliability indices without considering RES

In case I, the effect of the daily load curve, common mode failures, and planned outages is investigated on the reliability indices. In the first scenario (**S.1**), the test system is studied in the normal state without any of the expressed anomalies. In the second scenario (**S.2**), reliability indices are calculated by considering the daily load curve (load states). In the third scenario (**S.3**), changes of each index are investigated in the presence of the common mode failures for lines 113a/13b, 2a/02b, and 109/10. In the fourth scenario (**S.4**), the changes of each index are examined according to the intended maintenance plans for transformers T1011b, T1201b, and T0113.

Table 6.2 General load data

Load Name	Active Power (MW)	Reactive Power (MVAR)	Number of connected customers	Scaling factor for interruption cost	Load Name	Active Power (MW)	Reactive Power (MVAR)	Number of connected customers	Scaling factor for interruption cost
C10a	0.9	0.18	180	0.9	C22a	1.4	0.28	280	1.4
C10b	0.9	0.18	180	0.9	C22b	1.7	0.35	340	1.7
C10c	0.4	0.08	80	0.4	C23a	1.5	0.30	300	1.5
C11a	0.4	0.08	80	0.4	C2a	0.5	0.02	100	0.5
C11b	0.3	0.06	60	0.3	C2b	0.52	0.02	104	0.52
C11c	0.4	0.08	80	0.4	C2c	0.4	0.02	80	0.4
C12a	1.5	0.30	300	1.5	C2d	0.9	0.18	180	0.9
C12b	1.2	0.24	240	1.2	C3a	0.54	0.02	108	0.54
C12c	1.2	0.24	240	1.2	C3b	0.62	0.02	124	0.62
C13a	2.2	0.45	440	2.2	C3c	1.1	0.22	220	1.1
C13b	2.2	0.45	440	2.2	C4a	0.9	0.18	180	0.9
C14a	0.1	0.02	20	0.1	C4b	0.4	0.08	80	0.4
C14b	0.1	0.02	20	0.1	C4c	1.6	0.32	320	1.6
C14c	1.7	0.35	340	1.7	C5a	1	0.20	200	1
C15a	1.8	0.37	360	1.8	C5b	1.4	0.28	280	1.4
C15b	0.7	0.14	140	0.7	C5c	2.3	0.02	460	2.3
C15c	1.2	0.24	240	1.2	C6a	1	0.20	200	1
C16a	0.1	0.02	20	0.1	C6b	1.4	0.28	280	1.4
C16b	0.1	0.02	20	0.1	C6c	1.9	0.39	380	1.9
C16c	0.8	0.16	160	0.8	C7a	1.2	0.24	240	1.2
C17a	0.1	0.02	20	0.1	C7b	0.9	0.18	180	0.9
C17b	0.1	0.02	20	0.1	C7c	2.1	0.43	420	2.1
C17c	1.4	0.28	280	1.4	C8a	0.74	0.15	148	0.74
C20a	0.1	0.02	20	0.1	C8b	0.9	0.18	180	0.9
C20b	1.4	0.28	280	1.4	C9a	0.74	0.15	148	0.74
C21a	0.1	0.02	20	0.1	C9b	0.74	0.15	148	0.74
C21b	0.3	0.06	60	0.3	C9c	0.74	0.15	148	0.74
C21c	0.5	0.10	100	0.5	C9d	0.74	0.15	148	0.74

The intended maintenance plan for each transformer is as follows: **T1011b: 2–9 May; T1201b: 14–28 June; T0113: 10–20 November**.

The results for different scenarios and the key indices of the *Reliability Analysis* are presented in Table 6.5. As it can be observed in Table 6.5, the use of the daily load curve has had a significant impact on the SAIDI, CAIDI, and ENS indices. For example, the created clusters by "*Load states*" analysis for the load_C11a after applying the daily load curve are shown in Fig. 6.13. In addition, after applying the common mode failures, SAIFI and CAIFI indices had the most change compared to the base case (8.191% increase over the S.1). It should be noted that these indices are for the subscriber load point of view.

Table 6.3 Failure model for busbars, transformers, and lines

Busbar		
	11 kV	*33 kV*
Failure frequency for terminal	0.002	0.0025
Additional failure frequency per connection	0.005	0.015
Repair duration	15	24
Transformer		
Failure frequency	0.02	
Repair duration	343	
Line		
	11 kV	*33 kV*
Failure frequency (per km)	0.032	0.025
Repair duration	33.5	212

Table 6.4 Interruption cost for general loads in the form of time tariff

Interruption duration (min)	Interruption cost ($/kWh)
1	1.044
20	2.836
60	6.336
240	23.261
480	54.049
960	73.532
1440	111.678
2880	223
4320	394.456

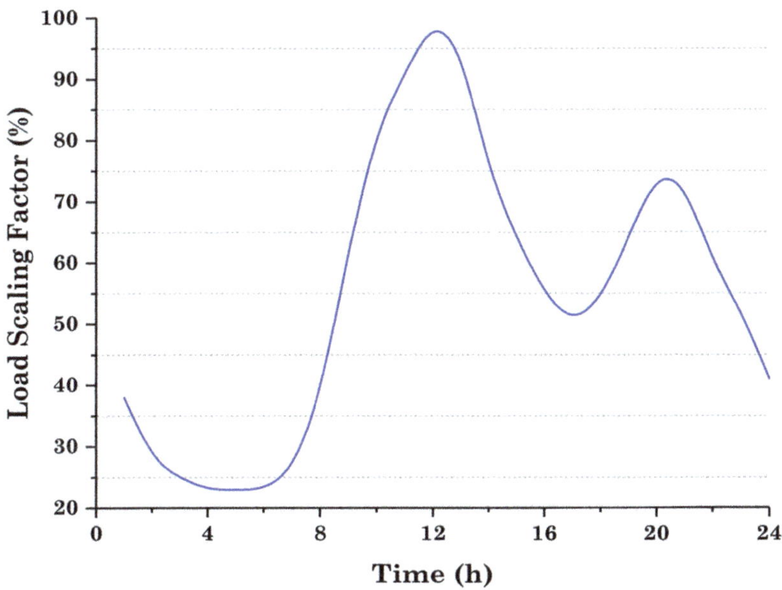

Fig. 6.11 The hourly load variations

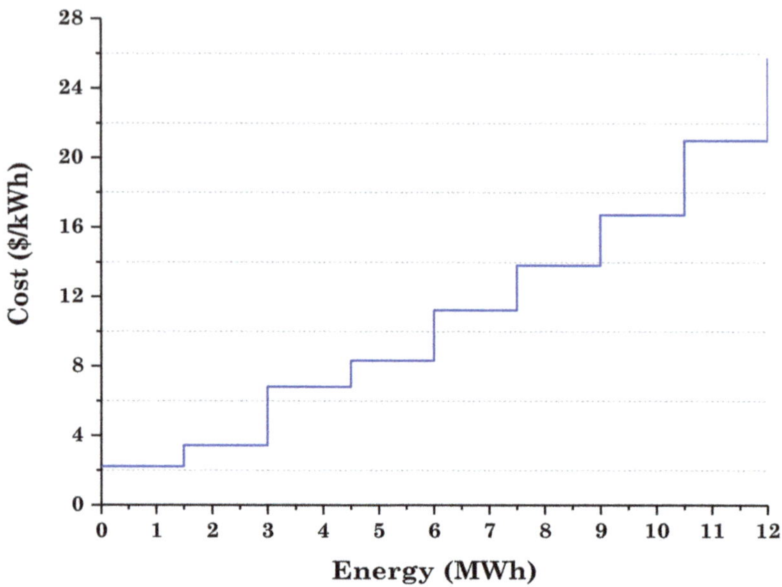

Fig. 6.12 The cost allocated to energy not supplied

Table 6.5 Comparison of technical indicators for different scenarios in case I

Indices	**S.1**	**S.2**	Changes compared to **S.1** (%)	**S.3**	Changes compared to **S.1** (%)	**S.4**	Changes compared to **S.1** (%)
SAIFI (1/Ca)	0.293	0.3	2.389	0.317	8.191	0.296	1.023
CAIFI (1/Ca)	0.293	0.3	2.389	0.317	8.191	0.296	1.023
SAIDI (h/Ca)	0.623	2.246	260.51	0.671	7.704	0.636	2.086
CAIDI (h)	2.124	7.485	252.4	2.112	−0.564	2.141	0.8
ASAI (%)	99.99	99.97	−0.02	99.99	0	99.99	0
ENS (MWh/a)	22.18	95.64	331.043	23.91	7.769	22.65	2.095

Case II: Investigating the reliability indices in the presence of RES

In this case study, the impact of utilizing the RES on the reliability indices is investigated by relying on stochastic models. To this end, three wind turbines and three photovoltaic (PV) systems are considered as renewable sources and located at B04, B09, B13 (buses to install wind turbines), B05, B08, and B12 (buses to install PV systems). The maximum power capacity of wind turbines and PV systems is equal to 3.6 and 2.6 MW, respectively. The hourly output power of PV systems and

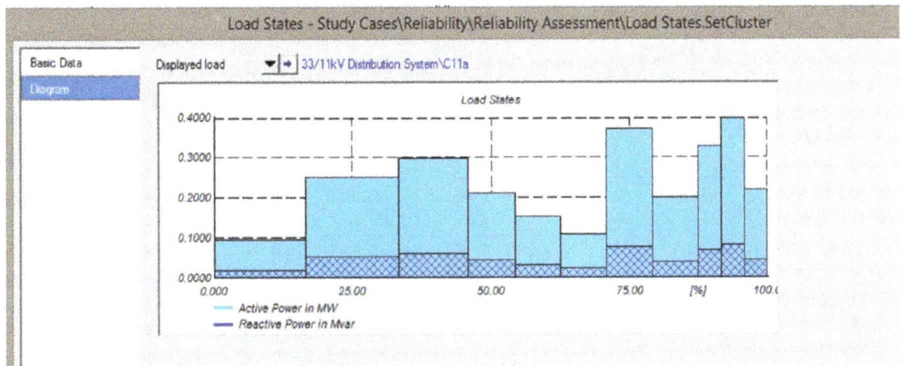

Fig. 6.13 The created clusters for load_C11a

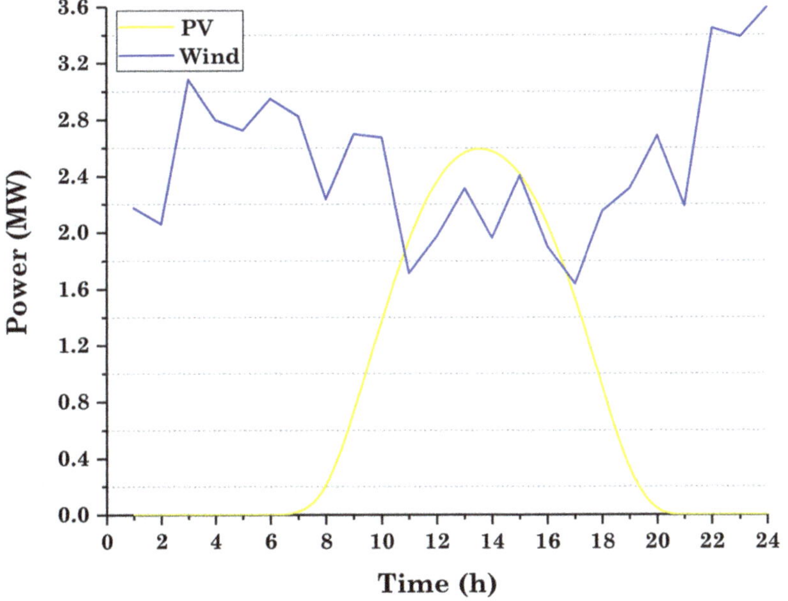

Fig. 6.14 PV systems and wind turbines power generation

wind turbines is shown in Fig. 6.14. It should be noted that how to use these curves in the PowerFactory is described in detail in Chap. 3. After installing renewable sources in the test system and applying the specified settings, *Reliability Analysis* must be performed for the introduced scenarios (S.1–S.4) in the presence of RES. The results for different scenarios in the presence of RES are given in Table 6.6. Table 6.6 shows that ENS index is drastically improved in the presence of RES. The best conditions in terms of improving the level of the system's reliability

Table 6.6 Reliability indices for different scenarios in case II

	S.1	Changes compared to C.I (%)	S.2	Changes compared to C.I (%)	S.3	Changes compared to C.I (%)	S.4	Changes compared to C.I (%)
SAIFI (1/Ca)	0.297	1.36	0.298	−0.66	0.322	1.57	0.301	1.68
CAIFI (1/Ca)	0.297	1.36	0.298	−0.66	0.322	1.57	0.301	1.68
SAIDI (h/Ca)	0.633	1.61	0.64	−71.51	0.681	1.49	0.642	0.94
CAIDI (h)	2.1	−1.13	2.147	−71.31	2.111	−0.047	2.127	−0.65
ASAI (%)	99.99	0	99.99	0.02	99.99	0	99.99	0
ENS (MWh/a)	7.87	−64.53	18.56	−80.59	8.472	−64.57	7.987	−64.74

are related to the use of daily load and power curves (S.2). In the second scenario, SAIFI and CAIFI indices are improved by 0.66%, SAIDI and CAIDI indices are improved by 71.5%, and ENS index is improved by 80.59%. The system operator can make appropriate decisions regarding the implementation of system development projects by using *Reliability Analysis* and evaluation of various indices under different conditions.

6.4 Calculation of Generation System Reliability Index by PowerFactory

The *Reliability Analysis* in the generation sector of power networks can be discussed in terms of system adequacy. Generating units differ from other elements such as busbars and transformers in terms of *Reliability Analysis*. Common problems that may arise for generation units in the power system include: (1) reducing the availability of generation units due to annual maintenance or various faults; (2) variations in renewable sources generation, which dramatically affects the production capacity in green networks; and (3) variations in system demand on an hour by hour basis. In PowerFactory, the *Generation Adequacy Analysis* is the best tool to evaluate the reliability indices of generation units. *Generation Adequacy Analysis* allows users to determine the share of renewable electricity generation to overall system capacity as well as the loss of load probability (LOLP) and expected demand not supplied (EDNS) indices. These indicators are the best criteria for evaluating reliability in the electricity generating sector. The PowerFactory uses the Monte

Carlo method to calculate these indices. The required steps to execute generation adequacy are as follows:

I. **Prerequisite step for performing *Generation Adequacy* analysis**

Users must define the stochastic model from the "Generation Adequacy" tab for each of the grid generators as well as the renewable sources as described in Chap. 3. For each generation unit, the available generation capacity of maximum output (in %) should be determined along with the probability of this availability (in %).

II. **Required settings to run *Generation Adequacy* analysis**

- *Press the "Calculation" button located on the main menu, and select "Generation Adequacy Analysis."*
- *Go to the "Initialize Generation Adequacy Analysis" option.*
- *After specifying the initial conditions, press the "Run Generation Adequacy Analysis" button and set the number of iteration to perform the "Generation Adequacy" analysis. Note that users can interrupt the analysis before the set number of iterations is complete by using the "Stop Generation Adequacy Analysis" button.*

The initial settings to the implementation of this analysis include four basic parts, which are expressed in Fig. 6.15.

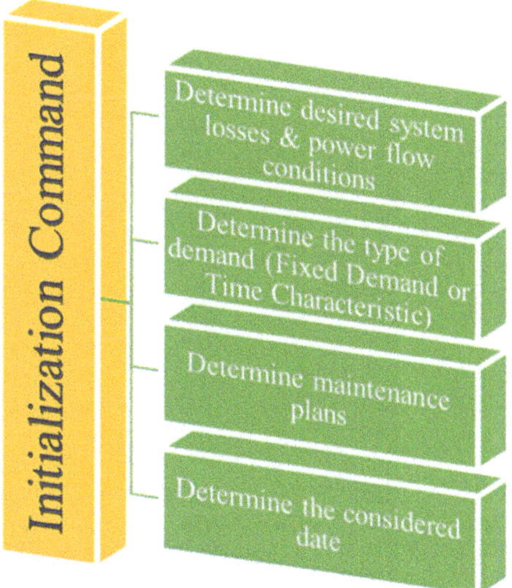

Fig. 6.15 Required initialization configurations in generation adequacy analysis

III. **View the results of the analysis**

LOLP and EDNS indices can be seen in the output window using the "*Output Calculation Analysis*" command. In addition to calculating these indices, this analysis provides various results in the form of distribution diagrams, Monte-Carlo diagrams, and convergence diagrams. After the successful implementation of the simulation, these diagrams will automatically appear in the "Generation Adequacy" toolbox. All results can be extracted from these diagrams are summarized in Table 6.7.

It should be noted that the LOLP index could be also obtained from the intersection of the total demand curve and the available dispatchable capacity curve.***The intersection of two curves equal to LPLOLP = 1-LP***

Table 6.7 Generation Adequacy analysis results

Type of diagram	Obtained result
Distribution plots	Total available capacity in MW
	Available dispatchable capacity in MW
	Available non-dispatchable capacity in MW
	Total generation (unconstrained) in MW
	Dispatchable generation (unconstrained) in MW
	Non-dispatchable generation (unconstrained) in MW
	Total reserve generation (unconstrained) in MW
	Reserve dispatchable generation (unconstrained) in MW
	Total demand in MW
	Demand supplied (unconstrained) in MW
	Demand not supplied (unconstrained) in MW
	Residual demand (unconstrained) in MW
Monte-Carlo draw plots and convergence plots	Loss of load probability (unconstrained), average in %
	Loss of load probability (unconstrained), lower confidence levels in %
	Loss of load probability (unconstrained), upper confidence levels in %
	Demand not supplied (unconstrained), average in %
	Demand not supplied (unconstrained), lower confidence levels in %
	Demand not supplied (unconstrained), upper confidence levels in %

6.4.1 Evaluation of LOLP and EDNS Indices

The introduced test system in Sect. 6.3.1 is used to analyze the generation adequacy of renewable sources. The availability of wind and photovoltaic units at different levels and the probability of each level are presented in Table 6.8. *Generation Adequacy Analysis* for the system under study is evaluated in various scenarios. The scenarios are as follows:

First scenario (S.1): For the first scenario, the test system did not utilize load characteristics and hourly power produced from RES. Hence, the *generation adequacy analysis* performs in the fixed load mode.

Second scenario (S.2): For the second scenario, the hourly load profile, as well as hourly power produced from RES is considered based on Figs. 6.11 and 6.14.

Third scenario (S.3): For this scenario, the Weibull distribution function in fixed load mode (like the first scenario) is used to analyze the generation adequacy indices. In this scenario, mean and beta coefficients are equal to 8.862 and 2, respectively, as well as the wind speed is modeled according to the presented curve in Fig. 6.16.

Fourth scenario (S.4): This scenario is exactly similar to the third scenario and also the effect of the Meteo Station Correlation on the LOLP and EDNS indices is investigated in the two modes (correlation #12 and #23).

Fifth scenario (S.5): In the fifth scenario, the generation adequacy simulation is performed according to the planned outages for transformers T0911 and T0708. It is assumed that the transformer no. T0911 will be repaired from May 10 to May 20. Moreover, the transformer no. T0708 will be repaired from July 1 to July 15. It should be noted that before defining new planned outages for the network's elements, previously defined maintenance plans should be removed. To this end, the following steps should be taken:

Data manager → Active project's library → Operational library → Outages folder → Remove previously defined maintenance plans.

The generation adequacy simulation results for each scenario are summarized in Tables 6.9. The presented results in this table illustrate the importance of modeling

Table 6.8 Stochastic model for renewable sources power generation

Photovoltaic systems			Wind turbines		
State	Availability (%)	Probability (%)	State	Availability (%)	Probability (%)
State 1	100	60	State 1	100	55
State 2	80	25	State 2	85	25
State 3	65	10	State 3	70	10
State 4	20	5	State 4	55	5
			State 5	30	5

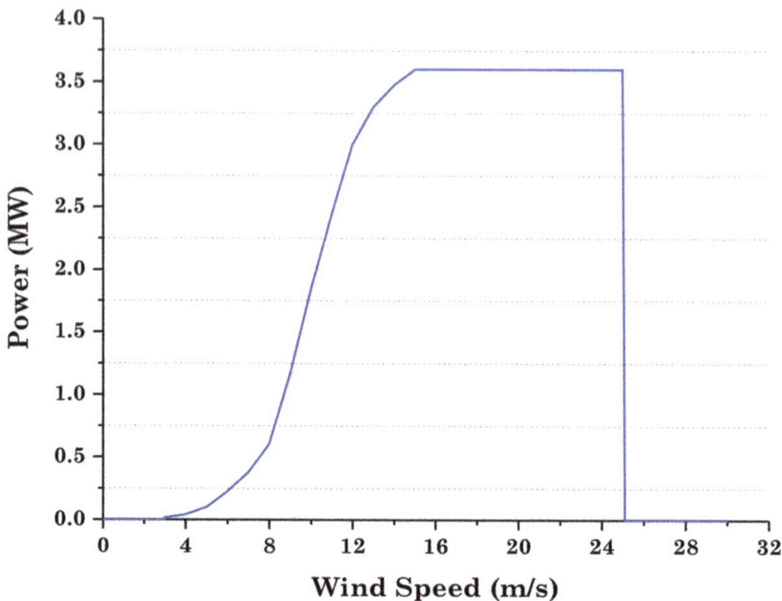

Fig. 6.16 Wind turbine power-speed curve

Table 6.9 Comparison of LOLP and EDNS indicators for different scenarios

Index	S.1	S.2	S.3	S.4 Correlation (12)	S.4 Correlation (23)	S.5 T0911/T0708
LOLP (%)	45.769	84.843	88.209	73.592	73.763	73.834
EDNS (MW)	0.564	16.217	2.901	3.394	3.391	3.397

renewable sources in determining LOLP and EDNS indices. As it can be observed in Table 6.9, the S.3 is the worst case in terms of the LOLP index. Therefore, the operation of power grids under different conditions can provide an engineering and economic perspective for the network operator to provide load demand in the proper situation.

References

1. Khalili T, Jafari A, Mohammadi-Ivatloo B et al (2016) Optimal battery technology selection and incentive-based demand response program utilization for reliability improvement of an insular microgrid. Energy 162:92–104
2. Yousefi-khangah B, Ghassemzadeh S, Mohammadi-Ivatloo B et al (2017) Short-term scheduling problem in smart grid considering reliability improvement in bad weather conditions. IET Gener Transm Distrib 11(10):2521–2533

3. Urgun D, Singh C (2019) A hybrid Monte Carlo simulation and multi label classification method for composite system reliability evaluation. IEEE Trans Power Syst 34(2):908–917
4. Zamani-Gargari M, Kalavani F, Mohammadi-Ivatloo B et al (2018) Reliability assessment of generating systems containing wind power and air separation unit with cryogenic energy storage. J Energy Storage 16:116–124
5. Suthapanun C, Jirapong P, Bunchoo P et al (2015) Reliability assessment tool for radial and loop distribution systems using DIgSILENT PowerFactory. Paper presented at the 12th international conference on electrical engineering/electronics, computer, telecommunications and information technology (ECTI-CON), Hua Hin
6. Sabpayakom N, Sirisumrannukul S (2016) Power losses reduction and reliability improvement in distribution system with very small power producers. Energy Procedia 100:388–395
7. Kamala S, Reddy BD, Sen B et al (2018) Improvement of power quality and reliability in the distribution system of petrochemical plants using active power filters. Paper presented at the IEEE international conference on industrial technology (ICIT), Lyon
8. Astapov V, Trashchenkov S (2017) Design and reliability evaluation of standalone microgrid. Paper presented at the 18th international scientific conference on electric power engineering (EPE), Kouty nad Desnou
9. IEEE Std P1366 -2003 (Revision of IEEE Std P1366-1998) (2014) IEEE guide for electric power distribution reliability indices. IEEE
10. Billinton R, Billinton J (1989) Distribution system reliability indices. IEEE Trans Power Del 4(1):561–568
11. Wang S, Liu L, Wu L et al (2018) Consumer-centric spatiotemporal reliability assessment and compensation model with sensitive component analysis in smart grid. IEEE Trans Power Syst 33(2):2155–2164
12. AlOwaifeer M, AlMuhaini M (2017) Load priority modeling for smart service restoration. Can J Elect Comput E 40(3):217–228

Chapter 7
Evaluation of the Real Grid-Connected Photovoltaic Systems in Iran

The main goal of this chapter is to investigate the effects of the real grid-connected photovoltaic (PV) systems on the distribution network. The networks used in previous chapters were ideal, so the impacts of renewable energy sources (RES) on improving the technical aspects of those networks were not well demonstrated. Therefore, a real test system, which is part of the East Azerbaijan distribution network in Iran, has been used to overcome this issue. The introduced tools in previous chapters have been used to perform the required analyses for evaluating the performance of the power grid in the presence of the real grid-connected PV systems. The general procedure for the **technical evaluation** of grid-connected RES is presented in Fig. 7.1. In the following, all the required steps to conduct a comprehensive technical study for grid-connected renewable energy systems are fully described.

7.1 Network Modeling

The unique capabilities of Geographic Information System (GIS) software can be used to model the selected distribution network located in East Azerbaijan. Hence, the development of an interface between the DIgSILENT PowerFactory and GIS database is crucial to perform the desired analyses. As shown in Fig. 7.2, after creating the standard interface between GIS and DIgSILENT PowerFactory, all data including network configuration are directly transferred into DIgSILENT PowerFactory from GIS. The selected distribution network is connected to the "*Golestan*" medium voltage (MV) substation and consists of 11 low voltage (LV) feeders with different lengths and subscribers. The information related to the LV feeders is given in Table 7.1. After network modeling in the DIgSILENT, we can use Load Flow analysis to evaluate the initial state of the various feeders. For example, the voltage profiles for feeders 1, 9, 10, and 11 in low-demand mode as well as without PV systems are shown in Figs. 7.3, 7.4, 7.5, and 7.6.

M. Zare Oskouei, B. Mohammadi-Ivatloo, *Integration of Renewable Energy Sources Into the Power Grid Through PowerFactory*, Power Systems,
https://doi.org/10.1007/978-3-030-44376-4_7

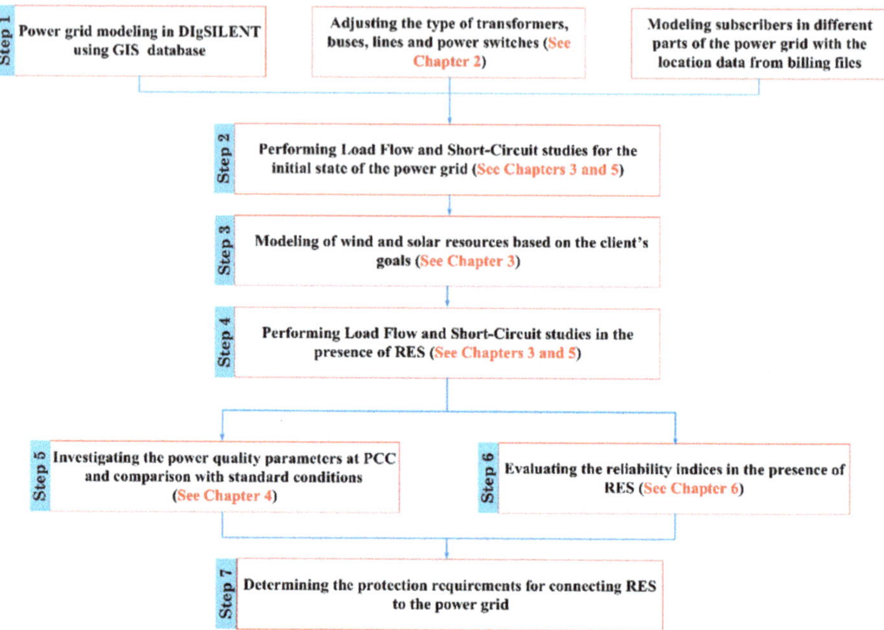

Fig. 7.1 Required steps for the technical evaluation of grid-connected RES

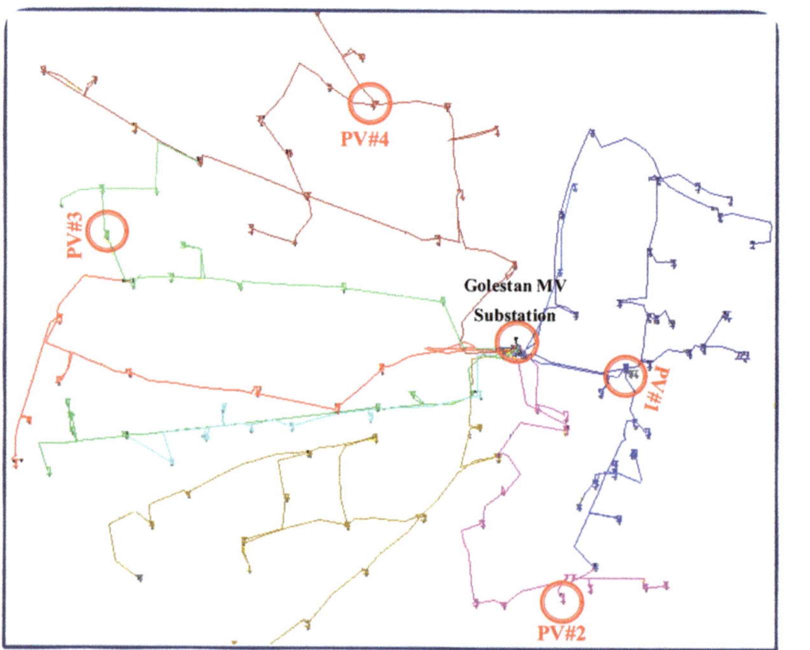

Fig. 7.2 Electrical network diagram of the selected distribution system in DIgSILENT

Table 7.1 LV feeders data

Name	Length (km)	Total active demand in (kW)		Total reactive demand (kvar)	
		Low-demand mode	High-demand mode	Low-demand mode	High-demand mode
F1_ Golestan	3.116	9707.85	12620.205	3828.6	4594.32
F2_ Golestan	1.066	2924.668	3802.068	1155.237	1386.28
F3_ Golestan	2.911	4680.977	6085.27	1846.056	2215.267
F4_ Golestan	4.655	5436.45	7067.385	2143.8	2572.56
F5_ Golestan	3.457	1651.65	2147.145	614.45	737.34
F6_ Golestan	0.625	627.75	816.075	248.4	298.08
F7_ Golestan	2.712	6575.955	8548.741	2596.249	3115.499
F8_ Golestan	3.931	4706.549	6118.513	1857.398	2228.878
F9_ Golestan	3.412	7553.635	9819.725	2974.694	3569.633
F10_ Golestan	4.154	7421.25	9647.625	2930.963	3517.156
F11_ Golestan	4.629	7169.643	9320.535	2828.901	3394.681

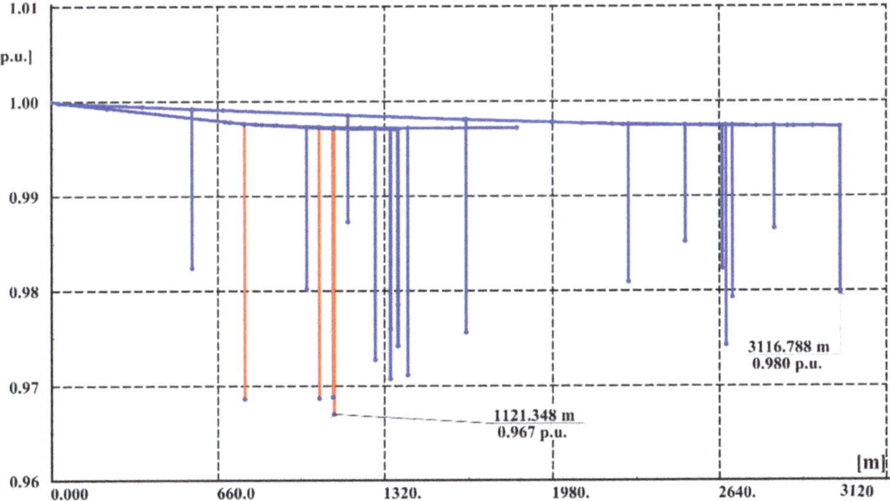

Fig. 7.3 Voltage profile for feeder "F1_Golestan" in low-demand mode without PV systems

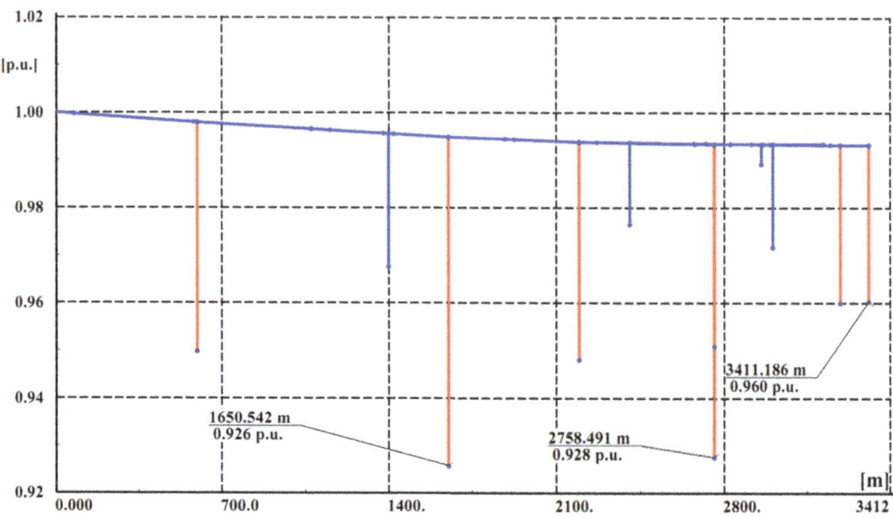

Fig. 7.4 Voltage profile for feeder "F9_Golestan" in low-demand mode without PV systems

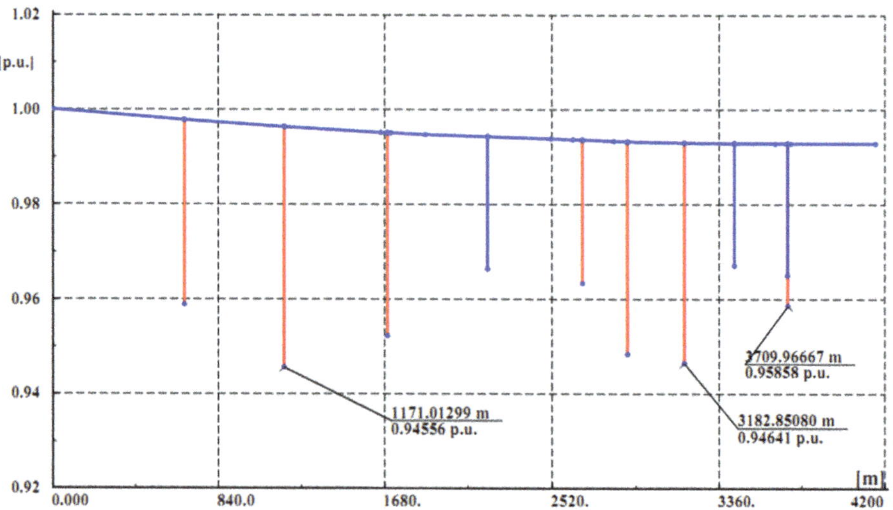

Fig. 7.5 Voltage profile for feeder "F10_Golestan" in low-demand mode without PV systems

After performing the feasibility studies by the distribution network operator, it has been decided to install four PV systems. The PV systems will be located at the specified points in the feeders 1, 9, 10, and 11. These points are shown in Fig. 7.2. The nominal power capacity of PV systems is assumed to be 20 kW (PV#1), 25 kW (PV#2), 25 kW (PV#3), 30 kW (PV#4).

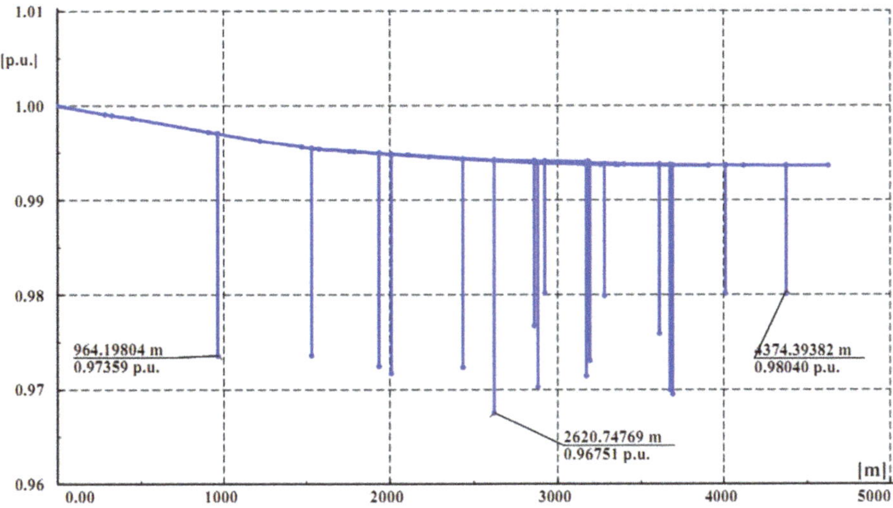

Fig. 7.6 Voltage profile for feeder "F11_Golestan" in low-demand mode without PV systems

Table 7.2 PV module specifications at standard test conditions

Electrical Characteristics					
Efficiency	Peak power	Rated voltage	Rated current	Open circuit voltage	Short circuit current
19.3%	315 W	54.7 V	5.76 A	64.6 V	6.14 A

Temperature coefficients of (%/°C)		
Power	Voltage	Current
−0.38	−0.27	+0.057

Correction factors for irradiance		
Solar radiation (W/m²)	Voltage correction (V/V)	Current correction (A/A)
1000	1	1
800	0.98	0.79
600	0.97	0.59
200	0.91	0.21

The electric power distribution company of East Azerbaijan has used Mono-crystalline, Sun-Power 315-SOLAR PANEL for the installed PV systems. The maximum power capacity and the efficiency of the selected panel are equal to 19.3% and 315 W, respectively. In Table 7.2, a summary of some main characteristics of these panels at standard test conditions is presented. The unique capabilities of PVsyst software can be used to perform comprehensive evaluations on economic feasibility analysis of the installed PV systems from the monthly energy yield perspective. PVsyst software has a wide database of meteorological data for various locations, PV system components and different manufacturers. In addition, this software simulates the performance of the PV system, taking into consideration the various possible losses.

7.2 Technical Analysis

In this section, we will evaluate some of the important parameters of the selected distribution network with/without considering PV systems using the various toolbox in the DIgSILENT.

7.2.1 *Load Flow Analysis*

Initially, active power losses, voltage drop, and maximum loading in each feeder are calculated for different modes through the *Load Flow Analysis*. The obtained results are shown in Table 7.3.

As it can be seen from Table 7.3, the total active power losses in both low and high-demand modes are reduced after applying all PV systems. When the distribution system operator uses PV systems, results show that the total active power losses in high-demand mode (3.21%) and low-demand mode (2.04%) are decreased in comparison with the initial status. In addition, the comprehensive results for the maximum lines loading in each feeder are presented in Table 7.3. As presented in this table, in the presence of PV systems, maximum lines loading in all feeders are decreased in comparison with the initial status. In terms of busbars' voltage levels, the worst conditions have occurred on feeder 9 in high-demand mode. By connecting PV systems to the distribution system, the status of this feeder is improved somewhat. For example, the voltage profiles for feeders 1 and 9 in low-demand mode as well as in the presence of the PV systems are shown in Figs. 7.7 and 7.8.

7.2.2 *Short-Circuit Analysis*

In the next step, the impact of PV systems on the short-circuit power level and also the short-circuit current is evaluated via the *Short-Circuit Analysis* for different modes. In this part, we evaluate the *Short-Circuit Analysis* for both three-phase fault and single-phase fault types. Table 7.4 presents the results of performing this analysis for several busbars, Bus_186, Bus_373, Bus_267, and Bus_369 during the three-phase fault and single-phase fault at all busbars. As shown in Table 7.4, the short-circuit power level and short-circuit current in the various status (with/without considering PV systems) are not much different.

Table 7.3 Comparison of obtained parameters through *Load Flow* analysis under different conditions

	Without PV systems						With PV systems					
	Low-demand mode			High-demand mode			Low-demand mode			High-demand mode		
Feeder Name	Max. loading (%)	Min. voltage (pu)	Losses (MW)	Max. loading (%)	Min. voltage (pu)	Losses (MW)	Max. loading (%)	Min. voltage (pu)	Losses (MW)	Max. loading (%)	Min. voltage (pu)	Losses (MW)
F1	85.94	0.967	0.174	112.74	0.956	0.266	81.50	0.969	0.163	105.11	0.96	0.242
F2	166.41	0.938	0.075	166.41	0.938	0.075	164.04	0.939	0.075	164.04	0.939	0.075
F3	180.85	0.928	0.119	180.85	0.928	0.119	173.97	0.931	0.117	173.96	0.931	0.117
F4	49.45	0.972	0.094	64.6	0.964	0.142	47.47	0.973	0.089	61.26	0.966	0.132
F5	78.17	0.972	0.031	102.33	0.963	0.045	74.75	0.973	0.029	96.48	0.965	0.042
F6	48.88	0.982	0.007	63.65	0.977	0.011	47.51	0.983	0.007	61.33	0.978	0.01
F7	94.14	0.963	0.124	94.14	0.963	0.124	93.18	0.963	0.124	93.18	0.963	0.124
F8	105.71	0.96	0.093	105.71	0.96	0.093	104.55	0.96	0.093	104.55	0.96	0.093
F9	189.31	0.926	0.246	189.31	0.926	0.246	183.34	0.928	0.243	183.33	0.928	0.243
F10	141.46	0.946	0.179	141.46	0.946	0.179	139.64	0.946	0.178	139.64	0.946	0.178
F11	75.45	0.968	0.13	75.46	0.968	0.13	75.04	0.968	0.13	75.04	0.968	0.13
Total active power losses (MW)			**1.273**			**1.431**			**1.247**			**1.385**

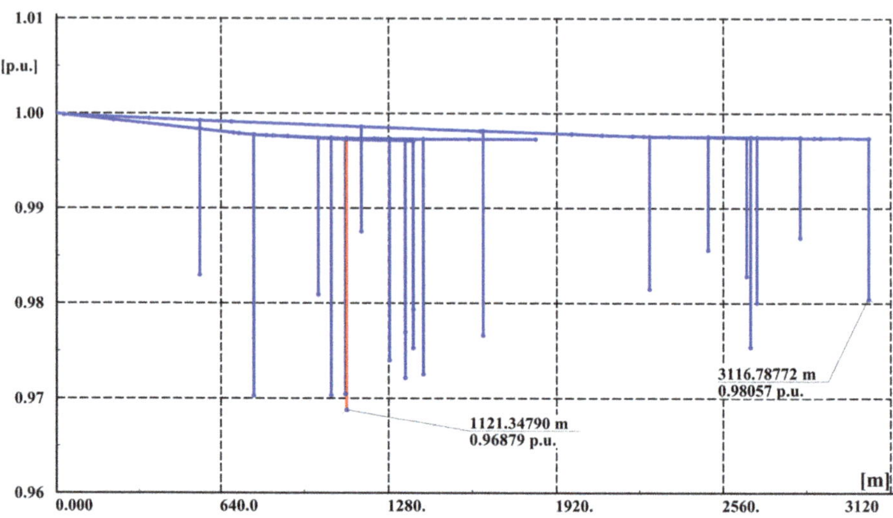

Fig. 7.7 Voltage profile for feeder "F1_Golestan" in low-demand mode with PV systems

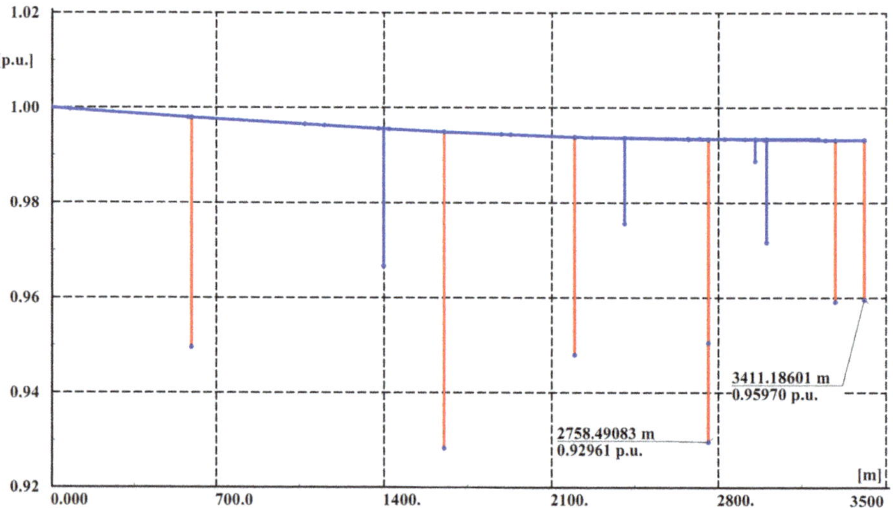

Fig. 7.8 Voltage profile for feeder "F9_Golestan" in low-demand mode with PV systems

7.2.3 *Harmonic Load Flow Analysis*

In this part, the total harmonic distortion (THD) index is investigated in the various modes. General loads and PV systems as the harmonic generation sources were modeled according to the available information in Table 7.5. The THD index at the point of common coupling (PCC) is calculated through the *Harmonic Load Flow Analysis* and the results of the analysis are presented in Table 7.6.

Table 7.4 Short-circuit current and short-circuit power level under different conditions

Bus name	Without PV systems				With PV systems			
	3PH		1PH (phase A with $R = 0.5\ \Omega$)		3PH		1PH (phase A with $R = 0.5\ \Omega$)	
	Sk (MVA)	Ik (kA)	Sk_A (MVA)	Ik_A (kA)	Sk (MVA)	Ik (kA)	Sk_A (MVA)	Ik_A (kA)
Bus_186 (F1)	1250.004	36.08	163.677	14.17	1250.024	36.09	163.676	14.17
Bus_373 (F9)	640.189	18.48	91.18	7.89	640.192	18.48	91.181	7.89
Bus_267 (F10)	599.631	17.31	86.04	7.45	599.634	17.31	86.04	7.45
Bus_369 (F11)	624.172	18.02	89.16	7.72	624.175	18.02	89.161	7.72

Table 7.5 Harmonic generation by PV systems and loads in balanced mode

Harmonic order	I_h/I_1 (%)	Phase (deg)
$h = 5$	24.05	0
$h = 7$	19.28	180
$h = 11$	15.09	0
$h = 13$	11.69	180
$h = 17$	9.88	0
$h = 19$	8.26	180
$h = 23$	5.34	0
$h = 25$	3.22	180

Table 7.6 Comparison of THD under different scenarios

Scenarios		THD at PCC (%)
Without PV systems	Low-demand mode	13.4
	High-demand mode	14.7
With PV systems	Low-demand mode	13.5
	High-demand mode	14.72

As it can be observed in Table 7.6, the THD index in the low-demand mode (with/without considering PV systems) has less value compared to high-demand mode. As expected, when using the PV systems in the distribution network the THD index is much higher than the initial mode.

Index

M. Zare Oskouei, B. Mohammadi-Ivatloo, *Integration of Renewable Energy Sources Into the Power Grid Through PowerFactory*, Power Systems,
https://doi.org/10.1007/978-3-030-44376-4

GPSR Compliance
The European Union's (EU) General Product Safety Regulation (GPSR) is a set of rules that requires consumer products to be safe and our obligations to ensure this.

If you have any concerns about our products, you can contact us on

ProductSafety@springernature.com

In case Publisher is established outside the EU, the EU authorized representative is:

Springer Nature Customer Service Center GmbH
Europaplatz 3
69115 Heidelberg, Germany

www.ingramcontent.com/pod-product-compliance
Ingram Content Group UK Ltd.
Pitfield, Milton Keynes, MK11 3LW, UK
UKHW021834270726
14058UKWH00001B/144

9783030443788